The ISTC Handbook of

Professional Communication

and

Information Design

Published by The Institute of Scientific and Technical Communicators:
First Floor
17 Church Walk
St Neots
Cambridgeshire
PE19 1JH

ISBN 0-9506459-5-8

Foreword

The past decade has seen a dramatic rise in the demand for good quality information about products and processes in industry, commerce and the public sector. At the same time, the rapid growth in the use of electronic documents has increased the range of careers open to professional communicators. There is a growing need for online training material, online help, and electronic documentation. Moreover, the rapid development of the Internet has led to an enormous demand for writers and designers of Internet pages and web sites.

The publisher of this book, the Institute of Scientific and Technical Communicators (ISTC), has been a support organisation for professional communicators for more than fifty years. In 2001 the ISTC are launching this book with the title 'Professional Communication and Information Design'. This title reflects the development of the communication profession and indicates the wide range of readers who will find the book useful.

The ISTC published its first book in 1985, the 'Handbook', which was revised in 1990. Some of the skills that were prevalent when the first edition of the Handbook was written have been superseded by the electronic age, but others, such as the need to communicate clearly, still survive. Technical communicators now write in fields as diverse as legal, financial, business and public information, as well as the more traditional technical fields. They write documents as varied as product information, training material, web sites, online documentation, and reference guides. Yet now, as in 1985, communicators must meet their readers' needs for information that is accessible and appropriate for its purpose.

Changes in the profession offer exciting opportunities for the communicator but also demand commitment and continuing professional development. The profession now requires a much wider range of knowledge and skills, and the communicator must continually acquire new knowledge and skills to meet the challenges of technological change and of the increasing public awareness of the importance of effective communication. This book offers a range of new papers by communicators, managers and academics in the United Kingdom, reflecting the range of knowledge and skills used by communicators today.

Professional communication is a challenging profession, and as professional communicators we have a particular social responsibility towards our readers who are dependent on the information we produce. Professional communicators must be sensitive to the needs of their audience, understand how to structure information and be able to use language and design to communicate information in an effective way.

By making a difference within their profession, they can take their part in maintaining the standards of professional communication that are just as important today as they were in 1985.

Contents

Introduction

In recent years the role of the professional writer has changed dramatically and in addition the profession has undergone a period of rapid growth. As a result, there is a need for professional communicators to share their knowledge and skills, and yet very few publications on this subject have been produced in the United Kingdom. This book aims to address this omission by offering papers which present a wide range of viewpoints on current issues in professional writing and information design. The papers in this book have been written by practitioners, managers and academics in the United Kingdom, and the book reflects their individual experience and opinions.

For many years there has been a debate about whether online documents will replace paper documents, and whether the paperless office is a viable reality. The book presents two different perspectives on this question. The issue is discussed by Dave Griffiths in the first chapter, which concludes that there will be an ongoing need for paper in some circumstances. Pete Greenfield, on the other hand, gives us an account of a paperless document distribution system currently in operation, describing the process by which the internal communications of the Abbey National Building Society became fully paperless, with the Society's communications being transmitted and updated by satellite.

These two papers show the importance of the audience and purpose of the documents in question. Kate Cooper likewise focuses on the needs of her audience when she addresses the problem of how to design an open learning course. This chapter also introduces some considerations more usually associated with hypertext, since the open learning course, although delivered on paper, had to be capable of being used in a non-linear fashion.

Nowadays the professional communicator may be writing in English for audiences of different cultures. Sandra Harrison asks how and to what extent we can adapt our writing to take account of readers from other cultures. This chapter looks at the strategies we can use when writing for different cultures, and includes references to a range of other resources.

Various stages within the writing process are discussed throughout the book. There is practical advice from Colin Battson on determining the audience and on creating a house style, from John Kirkman on writing style, from Ron Brown and Kathy Lawrence on editing, from Peter Lightfoot on illustration, from Richard Raper on indexing, and from Roy Handley on writing a synopsis. Increasingly, professional writers need to take control of the whole writing process from planning to delivery, and this challenge is addressed by Paul Warren in his chapter on managing documentation projects.

Newer forms of writing are also examined: Clyde Hatter explains how to develop a web site, Matthew Ellison discusses help systems design, and Bryan Little de-mystifies the subject of PDF files.

The appendices contain useful information about British Standards and proof-reading marks. The second appendix directs readers to the ISTC web site, which in turn opens up another exciting range of resources for professional and technical communicators.

For information about educational opportunities in technical authoring, contact the ISTC office, or see the ISTC web site Education page.

The ISTC would like to thank all our authors for their interesting and diverse papers, and also Sue Wood whose hard work and commitment have made this book a reality.

1

Design for usability - *no one reads manuals*

Dave Griffiths, MISTC *is currently working in the Information Design and Delivery unit in Kudos, one of the country's largest independent suppliers of Information Solutions. Before he joined Kudos in 1996, Dave managed ICL's largest in-house documentation team, known as Customer Information Solutions. Dave is a former president of the ISTC and is a frequent speaker at ISTC and other conferences.*

There are three ways of presenting my sub-title:

- **No one** reads manuals (*absolutely no one!*)
- No one reads **manuals** (*old hat technology*)
- No one **reads** manuals (*unless they have to*).

None of these statements is absolutely true. However, you have probably heard some or all of the sentiments expressed by friends and colleagues. A popular perception is that manuals are stored in bookcases, often in their original shrink-wrapping, that they have been replaced by Help files and internet browsers, and where they do exist, they are a refuge of last resort. But at least one philosopher has been quoted as claiming that 'perception is reality', so the manual designer does have to take these perceptions seriously. Manual design is not a simple task. Manuals are not the most popular products in the world. Manual design is neither a science nor an art form – it is both. Good manual design requires analysis of usability and creative solutions to particular situations.

Talking to users

Yet manuals can be good; in fact, they can be excellent. And they can be appreciated. When I managed a large documentation team for a major UK computer manufacturer, I placed a constant emphasis on authors talking directly to the end users of our documentation. One question I always asked was: 'Which manuals do you like and which do you dislike?' The answer to this question can be very revealing, especially if you find some commonality of answers from diverse users.

However, not all authors are enlightened enough to seek out the views of their readers. And poor examples of manuals are, sadly, easier to find than good ones. The most classic failure, in my experience, is the manual that has been written by an author who has assumed (perhaps subconsciously) that his reader will read the manual from chapter one to chapter ten. Yet for the most part when you speak to users, the message always comes back the same: 'We only look at the manuals when we are stuck'. When discussion continues, this turns out to be not quite the whole story. *Sections* of manuals are read, for example, to find out how to implement a new facility. But crunch time comes when the users are struggling to recover from an error – sometimes under time pressure and sometimes after in-context Help had failed to provide an explanation. On such occasions, being able to find the right section of text is equally as important as it is for the text to be understandable when found.

The purpose of this chapter is not to give advice on good writing, but to offer some advice on making the right information easy to find. It is one of the contentions of this chapter that manuals are rarely read through and this factor is of paramount importance for the manual designer.

Before delving into the main topic, it is important to set it into context. The title states that no one *reads* manuals. As I have mentioned this is, of course, a generalisation and not the whole truth. There are occasions when customers read our manuals – a product overview can reasonably be structured with the expectation that it will be read right through by a user who intends to gain an insight into the features of the product. There are many occasions where single sections, even chapters, of a manual will be read right through, by users who need to understand a particular topic, perhaps a newly acquired facility. For this type of usage, the manual designer still needs to take care that readers do not have to read Chapter 1 before they can understand Chapter 7. There are many ways to avoid this, some of which are covered here.

Moreover, it is also worth remembering that there are occasions where a manual that has *not* been designed to be read in a linear fashion is, in fact read in such a way. How many of you, on beginning a new job, have been told to get started by reading the manuals through? This is one of the activities that can cause manuals to give a poor impression of their true worth. And it is not the purpose of this paper to advise authors how to turn reference or procedural manuals into training manuals, but we do have to realise that we cannot determine the way in which our documentation is treated by customers.

Having recognised that some manuals are designed to be read through and that some manuals are read through even if they are not designed to be used in this way, my experience tells me that manuals are most commonly used in 'search' or 'look up' mode. A fundamental aspect of being a technical author is to be able to provide navigation techniques that help readers find the information they want, by whatever method *they* may choose in whatever particular environment *they* are likely to be in. Authors are not in absolute control of usability, but they do have a major impact on it.

How?

How do we do it? How can we design manuals to suit their anticipated (but not guaranteed) mode of use.

Well it's difficult to know where to start. Actually, no it isn't. You should start by talking to your users. Here is an example of an 'obvious' requirement that was initially missed by my own group of authors from the Customer Information project in ICL.

For many engineering systems or large software packages, the total documentation set rarely consists of a single manual. For example, ICL's mainframe operating system, VME, has over a hundred publications covering different, but overlapping, aspects of the system. So VME's documentation users do not just have the problem of finding information in a manual, they first have the problem of finding which manual it is in!

When Customer Information authors, as part of a programme that put authors in touch with their readers, visited VME sites across the UK, everyone they spoke to asked for a Master Index that covered all the VME manuals. In response to these comments the authors produced one, using a fairly painful process of merging all the individual manual indexes and then refining the total set. The resulting Master Index document immediately proved immensely popular. In reality the situation of a large suite of information existing without some form of global index should never have been allowed to happen.

However, when we considered just why a global index had not been included in the initial VME suite design, we realised that there was not such a large suite of manuals at the outset, and that the VME documentation suite had evolved as the operating system grew. This is a very important consideration – documentation design is rarely carried out with the proverbial blank sheet of paper. Documentation designs have to be constantly re-visited in order to adapt to changing situations. At this level of complexity a documentation architecture is needed to determine the structure of a complex suite of information and to set some rules for methods of implementation. But such an architecture, which is generally defined in a document of its own, must never be allowed to stagnate and prevent designers meeting the needs of their users in ever-changing situations. Incidentally, many manual design considerations also apply to suites of manuals as well as individual documents. Helping our users to find the information they need is a task that operates at many levels.

Whilst I have emphasised the need to talk directly to users about usability, I recognise that this is not always possible. There are two other common ways to measure usability (and hence influence design) – the use of tools and the use of in-house validation units.

There are many software tools that can provide a check on understandability. The fog factor index is the most well known. The fog index measures the mean length of words in sentences and the percentage of words with three or more syllables. However, it does not cater for different audiences, so you have to use its results with care. For checking the clarity of writing, I believe that nothing can compete with a 'second author' checking the text of the original author.

Commonly, in the computer application development environment, companies invest in permanent or temporary validation teams to hammer a new application alongside the manuals, which are generally at draft status at this stage.

This is an excellent process and probably the one that gives authors most of their useful input. However, the process does suffer from the fact that such validation teams are themselves made up of relative experts and they can overlook even fundamental errors in documentation, because they take so much knowledge for granted.

Structure

Fundamental design questions to be considered when structuring a manual for its appropriate usage include basics such as:

- What topics should be covered?
- How should they be linked?
- Should there be a linear structure or a hierarchical structure?

Typically a linear structure is suitable for some reference material (what better way is there to structure a glossary than by using the linear alphabet?), whereas a hierarchic structure is particularly well suited to task-oriented, procedural information. Manual suites are likely to contain both types of information structure. In fact, they may both exist inside one manual.

A facility may be organised into a number of sections, for example:

- A description
- A set of procedures or tasks
- A worked example
- A list of error conditions
- Interface specifications.

Or a topic may be so large that these components need whole manuals:

- A product overview
- A task-oriented procedure manual
- An error handling manual
- A command specifications manual
- A glossary of terms.

The essential point is to structure your material according to how your users will want to find it. You have to talk to your customers direct or, if this is not possible, talk to someone who knows how your users work. It is important to be certain to establish the structure right at the outset, when preparing a manual synopsis. Although today's publishing tools allow you to reorganise text with ease, once the first draft has been produced, an amazing amount of inertia seems to set in. If you find structural flaws at this stage the temptation is to patch the existing draft rather than redesign the document's structure.

Once the structure is determined, it is then imperative to be *consistent*. Use the same structure wherever possible throughout the manual and indeed throughout a suite of manuals. Customers quickly become used to a particular structure and will become adept at finding information if all the manuals have the same structure. I have known authors who have reorganised the structure of manuals to improve their usability only to receive comments back from users that 'I preferred the original'. The reality was that the customers were familiar with the original design and had learned to find their way around it. A balance has to be struck between allowing a design to disintegrate over time (particularly by the heinous tactic of adding extra chapters on to the end of a manual, because this causes less hassle for the author) and restructuring information to cope with changing situations. This concern is, of course, only relevant when you are documenting a product which itself has a long life. It obviously doesn't apply to manual editions that are never going to be updated.

One policy I have always advocated is the use of a section at the front of a manual entitled 'A note from the author'. In this section, an author can explain to readers what changes have been made and why. I personally prefer authors to sign such sections or even to include a photograph. This helps to create a direct rapport with your readers and should include a method for readers to feed comments back to authors. Many methods are possible, from simple fax numbers or email addresses to pre-printed tear-off forms. However, it has to be said that, in my experience, little feedback is actually elicited this way. Generally, you have to be much more proactive in seeking comment from readers. Telephone surveys or site visits are the most successful ways of polling reader opinions.

Navigation aids

Whilst structure is fundamental to the design of search mode manuals, there are many navigation aids that we can employ. The most obvious of these are the table of contents and the index. Yet there are many published manuals without one or other of these aids and some without either. This may be acceptable when the author wants to attempt to force a read-through approach onto a reader, but this is the exception rather than the rule and, as I have mentioned, you cannot control a reader's action. I would advocate that all manuals have both a table of contents and an index. Even in the case where, in a small manual, the topic entries may be mostly the same. Some users like indexes and some like to work from the contents. We have to cater for both types of user.

The production of good indexes is a topic in itself (covered in detail by Richard Raper in Chapter 8) so I will not delve further into this topic here. But I think it is worthwhile mentioning a few points about the construction of a table of contents. You do need to think carefully about the level of detail to include. Many word processing packages automate this process, but the designer needs to stay in control and override the automatic level of content generation if necessary. Simple chapter headings may well be all you require in an overview publication – you do not necessarily need to have every level of section heading.

For example, if you have a repetitive structure that looks something like this:

2.1 Use of Topic A
    2.1.1 Topic A in environment 1....................11
    2.1.2 Topic A in environment 2....................12
    2.1.3 Topic A in environment 3....................13
2.2 Use of Topic B
    2.2.1 Topic B in environment 1....................14
    2.2.2 Topic B in environment 2....................15
    2.2.3 Topic B in environment 3....................16

then this level of section heading is probably not particularly helpful to the reader.

If subsidiary sections are always small (and hence on the same page or next) you may choose not to slavishly include them in the table of contents if, by doing so, you detract from helping your reader find Topic A.

Whilst the table of contents and the index is fundamental to the usability of a manual, there are other search tools that you can, of course, use. Some of them, if not exactly unique to paper, are certainly most effective with paper manuals.

Page numbers are an obvious benefit of paper manuals. And headers and footers allow you to repeat section headings on each page. Protruding tabs and colour sections can all be used. However, don't use them unnecessarily. I have seen manuals of less than a hundred pages using six or seven protruding, coloured tabs that have actually made it difficult to turn to the page you require. I would generally recommend that tabs are only used for very large manuals and, if you do use them, please put a topic title on the tab. Don't use numbers or colour identifiers, because these require a two-stage look-up process for people who don't use the manual so regularly that they can remember that Section Turquoise covers turbo-charged widgets.

Within the body text, the standard methods of bulleted or numbered lists are, of course, useful devices. Less common, but certainly feasible, especially if there is a large print run, is colour highlighting. Simple spot colour is very effective and not too expensive if used with medium size print runs. I have seen colour highlighting used particularly effectively to show, for example, optional features of a product, which every user may not have. Marketing people often like this ploy, because it highlights optional features that the customer can buy.

As with structure, it is important to be consistent in your use of such navigation aids. The same techniques should be used throughout a suite of publications.

My own, personal, favourite aid to usability for technical manuals is not frequently seen - and I don't know why this is. It is the use of summaries or abstracts. A summary that states what you expect readers to be able to understand or achieve once they have read a particular section (or chapter) will show them immediately whether it is worthwhile reading that section or whether they need to look elsewhere in the manual. This device is something that many authors use in a synopsis to describe what topics will go into a manual, but is rarely carried forward into the manual.

Maintenance

Paper manuals are not more difficult to maintain than web sites. The difficulty is how to issue your updates to your customers. This section assumes that, for whatever reason, you are in the position of issuing printed copies of updates rather than PDF files or some other equivalent, electronic method.

The traditional way of updating paper manuals is to put them in ring binders and periodically issue an update with an amendment list of instructions:

- Replace page 19 by new page 19
- Insert new pages 3a and 3
- Replace Chapter 4 by new Chapter 4..........and so on.

I have yet to find a user who likes this method of update. Many users do apply them, because they understand the necessity, but I have also known users who file the update at the back of the binder to be applied when there is a 'spare moment' - *and there it stays!*

Exactly what is issued to customers may depend very much on the relative importance of the print budget, but I would advocate you apply some user-friendly rules to your update policy. So, for example, you should replace whole chapters rather than numerous individual pages within a chapter. This way you make the update much easier to apply and, therefore, increase the chances of it being applied. Whilst at ICL, I instigated a quality policy of replacing whole manuals if the update required technical changes to more than 10% of the pages. Once established, this fairly tight regime allowed us to move quite quickly the one step further to replacing manuals with new editions. Once this policy is adopted, you are, of course, freed from the restriction to use space-consuming ring binders.

My preferred form of binding for software and engineering manuals is Half Canadian Wiro binding, which combines the advantages of wiro binding mechanisms (allowing a user to fold a manual so there is only a single page footprint) with the ability to have a spine with space for a title that can be seen when the manuals are stacked in a bookcase or on a shelf.

The last thing to mention in this short section on maintenance is that users do like some indication of which areas of text have been changed from their current edition. The most common mechanism for paper manuals is sidelining. I find that almost all other methods such as highlighting changes, use of italics or different fonts, all detract from the general readability of the text.

Why paper?

Kudos was recently asked by a particular software house, for which it provides documentation services, to convert their system software documentation from delivery on a compact disc to delivery via the web. But prior to undertaking the task, we were requested to conduct a telephone survey of a sample of their customers to confirm that such a change would be acceptable. An interesting by-product of this survey showed that 25% of the users still retained the suite of paper manuals that preceded the delivery of the documentation via a browser on the CDROM.

Similar surveys have shown that many users prefer to print manuals if it is possible to do so from whatever electronic source is provided. Obviously this is not feasible for documentation that has been produced solely as context sensitive Help files or purely to be accessed in a browser environment via a search engine. However, many modern documentation systems are still based around a manual structure and PDF files designed to be printed at a customer's premises are a frequent method of distribution or as an optional accompaniment to hypertext-linked and/or search based HTML systems.

During a recent lecture I gave to technical communication students at Coventry University, I was asked if the use of web browsers and their as yet unidentified successors would ever completely replace manuals on paper. I don't *know* the answer to this question – nobody does – but I think the answer is no.

There are perhaps two reasons why many users favour paper manuals.

One is culture. People of my generation and older (that is, anyone who measures distances in feet and inches – and converts to centimetres only if they have to) grew up by learning from books, textbooks and manuals. They are our natural media for learning. For the new generations of the Information Technology age, the computer screen is the natural media for learning.

However, if this is the main reason why manuals remain in use, then their life cycle would be finite. And I don't believe this to be the case. However, I do believe that the use of manuals will become limited to specific areas where paper is actually the most appropriate medium for conveying the information. Some of the advantages of paper are probably subconscious even to the supporters of paper manuals, but they can be argued, even to the most vehement of on-screen fanatics.

So what are the real advantages of paper manuals?

Some things are fairly obvious:

- The user can see a large amount of information at one glance with paper. You can get more information on a printed page than in a screen
- Graphics can appear alongside the text it supports. Diagrams, charts and photographic work are significantly clearer on paper, and there is no need for pop-ups for labels and so on
- Users can annotate their copies with notes that are immediately visible when they turn to a page. This is possible electronically, but it is not nearly so easy to apply

- Paper is portable. So are laptops, but no one takes a laptop to bed (or do they?)
- Some people can absorb information more quickly from a page than from a screen. Perhaps this is linked to the culture category mentioned above.

Whilst most of this paper is the result of personal experience in the information dissemination business, some of my opinions are borne out by scientific evidence.

Some research work has been done into this topic by such luminaries as Pat Wright and James Hartley. Pat Wright found that readers of paper-based text benefited from the memory and place-keeping aids such as marginal notes, sub-headings, page numbers and particularly the ability to flick through pages.

I think it is sufficient to say in conclusion that an argument can be made for the retention of paper or indeed the selection of a paper manual structure (even if delivered electronically) for some new systems or applications. To come back to the point made earlier, the information designer needs to ascertain what delivery medium is appropriate to the intended readers in their environment.

Modern technology has provided us with a range of information delivery mechanisms. But I do believe that paper is not an obsolete method just because it is an old method.

2
Technical editing in ten easy steps

Ronald Brown, FISTC *has degrees from Brunel University and Imperial College, and leads a varied career as a researcher, academic and technical communicator. He has been a full time technical author since 1986, mostly as a team leader. Currently principal technical communicator at KBC Process Technology Ltd, he writes and edits paper and online documentation.*

Kathy Lawrence, *armed with a degree in Management Sciences from the University of Warwick, gained commercial experience at computer company ICL before moving into journalism. Now operating as Wrightwell Editorial Services, Kathy and her team provide writing and editing services to the marketing and PR departments of many blue chip companies.*

This chapter outlines the basic features of editing, and provides examples and tips on how to structure and lay out material, and how to check it. We expand on these points through examples, and touch on the many false starts that form the basis of our experience. Our text draws on terminology and ideas from our differing backgrounds in editing, technical authoring, publishing and copywriting , combined to provide a general guide on the role of the editor.

Introduction

This chapter deals with the task of editing other people's work. As well as a good understanding of language and an eye for detail, good editors need a degree in tact and diplomacy. They remember that authors may have laboured and sweated over every word, phrase and illustration, only to have to yield up their work for someone else to check. Authors may feel overprotective and proprietary about their work, and the last thing they want to hear is that it isn't quite right.

As a result, the editor–writer relationship can sometimes be cool, or occasionally even hostile. To quote E J Marden[1], 'In all probability, a dentist enjoys a better relationship with his patients than most editors do with their authors.'

But it doesn't have to be like this. We introduce the editor as a friend, explaining to author and editor what the role of the editor should be, and the added value that the editing process can give to a document. We aim to help develop the feeling that editor and author can work in partnership to improve and further strengthen the document. And the better the final product, the greater the esteem attached to it.

Style guides

It is normal practice in a book or journal publishing house or company that editors (and we hope authors) have a document explaining the housestyle. The style guide sets standards for the publisher's preferences on topics such as layout, spelling and conventions to follow. It is invaluable in giving a consistent look and feel to a publication – especially when it has been written by many authors.

If you are starting from scratch, you may not have such a document. This chapter outlines the areas you should think about as you start editing, and gives practical suggestions on what to watch out for, so you can develop your own style guide. If you're really organised, you can even distribute the guide to authors before they start writing, so they all work to the same specifications. This saves a great deal of time later, for everybody.

Role of an editor

What does an editor do? There are three main tasks, which often overlap.

- **Organisation and structure**
  Ideally, you consider the organisation and structure of a document in advance, with editor and author agreeing on how the document will be developed and the subject matter that it will contain. Author and editor can continue to liaise as the document is written.
  Once it's completed, the editor needs to look at the substance of the work – such things as overall contents, meaning and possible omissions, rather than details of language and construction. You want to improve the coverage as a whole.
  At this stage you might consider the following points. Are there any major gaps, or areas where the content isn't right? Does the document contain unneeded material? Is the sequence right?
  This is probably the place to watch out for legal requirements. You need to make sure the document uses trademarks correctly, is copyrighted, and acknowledges any material whose copyright belongs to someone else.
  This is sometimes called *substantive*, *developmental* or *comprehensive* editing, in case you start to read other books on the subject. For convenience, we apply the term *commissioning editor* (commonly used in book publishing) to someone working at this level, who has overall responsibility for the document.

- **Copy editing**
  This involves detailed editing for sense, correct use of language and spelling, looking for contradictions and inconsistencies.
  All of these tasks apply to just about any kind of copy editing. But there are some additional functions for technical editors:

  - **Technical accuracy**

    You must have a reasonable background knowledge in the subject area, or have access to a subject matter expert.
  - **Level**

    Part of the editing task is to make sure the technical content is appropriate for the intended readership.
    For convenience, we apply the term *copy editor* to someone working at this level.

- **Proofreading**
  This checks that material is presented clearly and cleanly, ready for printing (paper or electronic), and that no errors have crept in during the production process.

The overlap between the various types of editing occurs because sometimes you can't see the general structure when a document is cluttered with inaccuracies or inconsistencies. Equally, sometimes when you have sorted out the details, you realise that there are major gaps, or that the material is jumbled. Additionally, there can be an overlap when one person does all of the editing tasks.

Background

As a copy editor, you need to know what the end product is expected to look like. When you're collecting together the material to be edited, it is always a good idea to ask for a sample of previous work. This often answers your queries, without having to refer back to the commissioning editor.

If you have doubts or questions, contact the commissioning editor, or the person who has overall responsibility for the project. If you are editing a technical author's work, you may be able to talk to them directly. This helps to avoid wasted effort on your part going down the wrong path, and lessens the offence caused to the author by major editing.

Remember that you are editing a document to conform to a certain style and for correctness, not to your particular taste.

Paper and online editing

Traditionally, editors worked with paper, and used proof-reading marks[2] that were more-or-less standard, and were clearly understood by editor, author and typesetter. Nowadays authors have essentially become their own typesetters, but proof marks still allow editor and author to explain to each other the changes they want to make. An understanding of these marks continues to be very useful, and Appendix 3 gives examples.

Alternatively, an editor can work online. This way, you can change the text directly on the screen. In word processing packages, you can highlight the changes with electronic markup. For example, in Word choose **Tools - Track Changes** to add proof marks electronically as you make changes.

There are other things you can do when working online. You can check the format using the template, for items such as styles, paragraphs and fonts. This way the editor can make sure an author has used the correct menu items and styles, or apply the correct ones if the author hasn't done this. In Help text, you can check the number of clicks required to get to the information you want. In Adobe Acrobat PDF (portable document format) files and in Web pages, you need to cross-check every link, to make sure it is active and takes you to the right place.

Whatever the mechanics of the process, the task and the skills required are much the same.

Skills required

What skills does an editor need?

At the commissioning editor stage, you must be able to see the overall flow and aims of the document, be able to analyse and evaluate the content, and have the ability to put yourself inside the mind of the reader. An editor acts as the intermediate between author and reader, and sees things from both points of view.

At the copy editor stage, you must have a good understanding of language and its rules, know the grammar, and have a good vocabulary. You must be able to concentrate on the text, and focus on detail. You must have a good eye for layout, and be able to spot inadvertent font changes. You must also understand the technical content, or have access to a subject matter expert.

In short, to edit you need the whole body of knowledge and skills that form the basic toolset for the theory and practice of professional technical communication.

All of these are skills you can gain by actually doing the job.

What is the result?

The main contributions an editor can make are on the

- impact
- readability, and
- effectiveness

of the document. When it works well, the joint efforts of author and editor can even help to produce something that is both easy to read and interesting, or perhaps even stimulating.

Ten easy steps

Having placed the role of the editor in context, we now describe the actual editing process. We use ten steps to illustrate the fundamental tasks that an editor has to carry out, providing you with enough information to begin the editing process.

We look first at how you need to consider the document as a whole, then focus on the details.

1 Overall impression

Since we get an overall impression of a piece just from the way it looks, part of the impact of a document comes before you get as far as reading the first word. The kind of things that contribute to the overall look and feel include:

- Is there a large mass of black text unrelieved by illustrations, headings, bullet points or numbered lists?
- Do text and figures work well together? Do the illustrations suit the intended audience? For example, graphs usually work best in highly technical manuals, but pie or bar charts might be better in more general documents, such as annual reports

- Is the typeface suitable? Are lines closely packed with a small typeface, so they are difficult to read?
- Are equations and formulas well presented and attractive?
- Are there copious, difficult looking mathematical equations, or is data put into graphs and tables?

A good layout is simple, uncluttered, straightforward and easy to follow.

If a publication looks good, it probably is good. At least, we start with a feeling that it's good, and then have to be persuaded otherwise. If a publication looks bad, it probably is bad.

2 Structure

Technical documents usually deal with large amounts of often complex information. It is essential to present the information clearly, simply and logically, so it is easy to follow and understand. A good structure helps.

A document with good structure make sense, follows a logical order and flows well. Check on the most important aspects of the information, and see if they are obvious and easy to spot. Look at how the material is organised. Are the topics self consistent and complete? Is material on one topic scattered about in different places?

So how can you improve the structure?

- Make sure the document follows a logical, coherent, well-thought-out sequence
- Make sure ideas, results and topics are grouped, so the reader can see both pattern and purpose in the sequence
- Check figures are near the text that goes with them, and are in the right order
- See if you can break very long, complex chapters or sections into smaller ones
- Select material, and put related subject matter into the same section
- See if you can put highly detailed or peripheral information in an appendix.

Poor choice of words, bad style, incorrect grammar and poor presentation may leave the reader struggling to understand the meaning. But disorganised, thoroughly mixed material may be so obscure and confusing that it is impossible to understand it at all.

The structure of a technical document depends on the topic, type of publication, intended audience and where it is due to appear. It is the editor's job to make sure the structure is appropriate for its intended use.

3 Clarity

Probably the single most important aspect of a technical document is that it says exactly what the author means. Too often we identify obscurity with academic excellence or status. Sometimes, we think that obscurity and ambiguity are unavoidable, simply because the subject matter is highly technical.

Many technical documents have to deal with complex topics that have their own very specific terminology. Most of the time it isn't these things that make the writing unclear, but bad presentation, poor style and not saying what you mean.

If the documents you edit are written by engineers, scientists, computer programmers or other highly skilled specialists, they will often use jargon, technical clichés and prefabricated constructions. John Baker[3] uses the term *grandiloquent writer* to describe this type of person. By this he means someone who uses abstract words and concepts where none is required, and who continually inflates the prose. It is the editor's job to make sure that, as far as possible, the document uses simple, direct language, and the meaning is clear.

If you are editing work that has been done by a technical author, then it should already be clear and to the point, since these are the basics of the trade. But even then, it is worthwhile asking: 'Does it say what it means and mean what it says?'.

Of course, it doesn't necessarily follow that all technical specialists produce unclear documents, or that all technical authors produce clear ones. As an editor, you have to be vigilant, whatever the source material or its origin.

It has been proved many times that writing in the active rather than the passive is far more interesting, and easier to follow. Technical documents tend to be written in the passive, as this sentence is. It may be difficult to persuade your author to adopt another style, and it's certainly not advisable to rewrite text to make it more dynamic and clear without the author's agreement.

George Herbert has a useful maxim: 'Good words are worth much and cost little.' This applies especially when you are checking for clarity.

4 Consistency

One of the fundamentals is to check for consistency. Make sure the document follows the same conventions for layout and usage throughout. Otherwise, it confuses the reader and makes it more difficult to hold their attention.

Check that the typeface, numbering and style for chapter headings, and heading levels 1, 2, 3 ... are consistent. Check that the headings are always in sequence, so heading 1 is always followed by heading 2, then heading 3 ... Check the typeface for text, and the layout of the contents page and index, if there are any. Check that headers and footers are consistent.

Be careful that pages are in sequence, and that odd and even pages are correct, especially in a chapterised book.

Other things to watch out for are:

- **Numbers and units**
  Check for consistency in numbers and units. Do you use a comma in a number or a space, such as 10,000 or 10 000? See that sentences don't start with a number, and look at how the text handles two, consecutive numbers. Technical authors often find it easier to communicate with words or pictures than with numbers[4], so it is worth checking these carefully.

- **Lists**
  Are lists numbered or bulleted? The convention is to number lists when you have to follow a sequence. Otherwise, use bullet points. Check secondary lists. Are lists consistent?

- **Tables and figures**
  Check that the numbering (if any) is consistent; check the style of numbering and the way tables and figures are referenced in the text. Check screen capture images and half-tones. Figures and tables (including their captions) need to complement, rather than repeat, information in the main body of text.

- **Bold or italics**
  These are often used for new concepts or unusual terms; italics are used in publication titles and foreign words. Check for consistency.

- **Capital letters**
  Check the use of capitals, especially initial capitals, to make sure it's consistent. If there are a lot of capital letters, suggest reducing them. Modern documentation usually minimises capital letters, which tend to clutter the text. But watch out for trademarks (described later).

- **Mathematical terms and equations**
  Many technical and scientific texts require the ability to write and use equations[5, 6] to the same level of competence as the rest of the document. When editing, make sure that the information in equations is not only easy to follow and well laid out, but is also accurate and correct. Note that the minus sign is en–dash and not a hyphen.
  Also check the way symbols appear, especially if the document was produced using a variety of desktop publishing packages. Sometimes, different word processing packages use different symbols in their character sets, and you can get odd substitutions.

- **Program listings and commands**
  If these appear in a different font, make sure it is consistent.

- **Footnotes**
  Are these always numbered in the same way, and do they always use the same font and style?
- **References and bibliography**
  Does the document use reference numbers within the text with a list at the end, or are the references put in with author and year? See how the references are laid out; every publisher has a style on this, and it is sometimes complicated.
- **Indexes**
  You need to watch for consistency in the use of words, spellings and capitalisation in the index. Often, an author changes things during the assignment, but may forget to change index markers. Sometimes, the index markers aren't part of the spell-checking. Also watch the index for conciseness. You may find several terms can be grouped under a level 1 heading and several level 2 headings; sometimes a word appears twice, once as a singular and then as a plural.

5 Conciseness

Why is this important? Unnecessary words cloud the meaning, and add to the time it takes to read the document. They also try the patience of the reader. Always try to remove unnecessary words. Also, check to see if you can use a good short word rather than a clumsy long one, provided the meaning is the same.

Technical documents tend to be overlong, rather than too short. Experience suggests that some authors are terrified of ending a sentence, let alone a paragraph. Help them to help their readers by making the document more concise and improving the flow. This is increasingly important when you are working on documents that have limited space, such as a brochure, where immediate impact is key.

Look at the following three edited examples:

1. These values are specified to a higher precision than …

   These values are more precise than …

2. An upper bound is placed on the conversion level, so as to keep the product stable.

   The upper bound placed on the conversion level keeps the product stable.
3. The values of the hot soak losses in the winter are explicitly defined to be zero.

   In the winter, the values of hot soak losses are set to zero.

Although the difference is only a few words each time, the concise sentences have a greater impact, and are easier to follow, than the longer ones.

Check figures and the text that goes with them. Often, an author tends to repeat information in a figure or its caption, and in the text.

Robert Frost has a useful saying. 'The more you say the less people remember.'

6 Spelling, grammar and language

If authors vary the way they spell words, make spelling mistakes and errors in grammar, or use inappropriate language, they distract and confuse the readers, and make it more difficult to keep their attention. People with a good grasp of the language are unaware that they mix rules, singulars and plurals, and genders with great abandon.

Even where spelling is correct, an issue that regularly arises is whether to use *–ise* or *–ize* endings. This is made more complicated when you have to localise documents for either a British or US market. If you see optimize and maximize, make sure the rest of the text is consistent and doesn't switch to realise and minimise. Watch out for words that always end in *–ise*, where the *–ise* is part of the stem, such as advertise, enterprise, supervise or televise. If the writer used a spell-checker effectively, this normally picks up most stray occurrences.

Make sure subject and verb agree. When they don't, it may be because the sentence is clumsy, or because there is room for doubt.

For example:

> A number of experiments *was/were* carried out ...

is the kind of construction that prompted letters in *The Times* a while ago. Less than half the people saw *number* as the subject and more than half saw *experiments.* If you can't decide on which is correct, rewrite the sentence to avoid ambiguity. The text then flows better, and the reader won't hesitate over the construction. In this example, one solution is:

> Experiments were carried out ...

Here is another case, this time where a competent author and editor couldn't agree:

> The mark plus descriptor/taglines is available on disc...

which ended up as

> The mark and descriptor/taglines are available on disc ...

When checking grammar, don't be too much of a purist. Language is not a fixed concept and rules that were imperative 50 years ago just don't matter so much now. Give an author some leeway with an occasional split infinitive (*The New Fowler's Modern English Usage* has a great deal to say on this) or other lapse of grammar, especially if it makes the text clearer and more lively. Allow for the artistry of the writing process.

For example, the following has an impact that would be diluted if the infinitive were not split:

> It is vital that all the documents we produce have a common look and feel, to further strengthen our corporate identity.

Make sure the language is right for the particular task. Whatever your views on political correctness, it is generally accepted that documents need to avoid emphasising gender. If the author hasn't considered it, you can usually find a generic word that doesn't depend on gender, or suggest rewriting a sentence if all else fails. Also suggest rewriting to avoid the *s/he* construction, which most readers find clumsy.

7 Punctuation

Why is punctuation so important? Because even a lowly comma out of place can completely change the sense. Consider the following:

> She slapped the face of the man who kissed her.
> She slapped the face of the man, who kissed her.

But a misplaced comma or full stop can do more than merely change the sense. There is a nerve wracking (for a technical editor) story in the *Financial Times*[7] about the Lockheed C-130J Hercules transport aircraft. In one of the company's sales contracts, a comma was misplaced by one decimal point in the equation that adjusted the cost for inflation. Lockheed estimates this one misplaced comma will cost it $70 million.

As part of the punctuation, check for consistency in the use of abbreviations, apostrophes, hyphens (-) and the en-dash (–) and em-dash (—), solidus, quote marks and parentheses.

To hyphenate or not to hyphenate is a question that can cause surprisingly heated debate. The important thing is to decide a rule and stick to it. For example, do you co-operate or cooperate? Is it email, e-mail or E-mail, and it's either CD ROM or CD-ROM. If the text has nonlinear, it shouldn't also have non-recurring; and you might have lower-case, lower case or lowercase, but not a mixture. This is complicated by the debate about compound adjectives. For example, if you use lower case, does that make it lower-case text or lower case text? The former makes the sentence easier to understand quickly and avoids misunderstanding, but using hyphens to combine phrases in this way can lead to a heavily hyphenated text (or is that heavily-hyphenated text?).

Make sure the apostrophe is kept for the possessive, and is not used in plurals. We have recently left the 1990s, not the 1990's; the plural is CDs not CD's, and similarly CD ROMs not CD ROM's. Watch out for the exception: *its* is the possessive and *it's* means *it is*.

Decide whether to use single or double quotes. Whichever you choose, use the other option for quotes within quotes. And if you're using a quote, do you put the full stop or comma at the end inside or outside the quote mark?

Other things to consider, all in the interest of clarity and brevity:

- Do you use full stops in acronyms and abbreviations, such as U.S.A. or USA, e.g. or eg (best not to – it's a waste of space and distracting)
- Do you accept Latin terms? It's normally best to avoid them, especially for an international audience
- Why use :- at the beginning of a list, when a simple colon will do?
- Do you accept ampersands, or replace them with *and*?

8 Trademarks and copyright

Trade names and trademarks identify a company and its products to the marketplace. Make sure a trademark is used correctly, exactly as it is registered. Companies spend a lot of effort developing trademarks for brand recognition, and a lot of money protecting their marks. A useful check is to look at a manual from a company, since this usually contains their trademarks (registered and unregistered) on the back of the title page, and gives the correct company name. You can then check the way a company is using its mark with the way the document you are editing is using the mark.

Most trade names have an initial capital letter, and companies often have a special way of writing them. AppleTalk and EtherTalk are registered trademarks of Apple Computer, Inc.; PowerPoint is a trademark of Microsoft Corporation; PageMaker and PostScript are trademarks of Adobe Systems Incorporated. It is UNIX or Unix, depending on who supplies the software.

Watch out for changes in company names and affiliations. For example, PageMaker used to belong to Aldus Corporation; BP is now BP Amoco.

Registered trademarks are often shown with an ® symbol when they are listed; unregistered ones often have a ™ symbol.

Check there is a copyright notice. This normally contains the © symbol, along with the word Copyright, the date of publication and the company name.

9 Queries

There will always be questions that you need to ask. Devise a system of making a note of them and what they relate to in the text, and make sure you get the chance to point them out. Make sure you get an answer, and deal with it. Perhaps produce forms that can be forwarded to the author.

10 Proofreading

In an ideal world, get a fresh pair of eyes to look over the copy once before it goes for final printing, especially if the document has been through further processing of layout and design.

Proofreading is the last check for consistency, accuracy and layout – it's the final quality control. Most documents take a while to appear, and go through several drafts. Usually, things change as the project progresses, and errors or inconsistencies can creep in. The proofreader has the final chance to correct these errors.

Proofreading is especially needed if the document is being transferred from one word processing document format into another. While this is not as major a task as it once was in the days when all text had to be keyed again by typesetters, it is still surprising how many typographical changes can happen to text when it's imported from one word processing program to another.

Never omit the proofreading task.

Endpiece

An editor is not the author's enemy. The editor can help the author bring consistency, clarity, completeness and polish to a document, ensuring that the document achieves its goal with the maximum effect. For the copy editor, this involves considering the document as a total entity and in detail.

This chapter has set out the main areas that an editor needs to consider, giving examples from our own experience. The further reading list provides you with a deeper understanding of the issues and practicalities involved.

References

1 Marden, E J (1985) *The ISTC Handbook of Technical Writing and Publication Techniques*, Chapter 5, 'Editing', page 100. William Heinemann, London.

2 See, for example, British Standards BS 5261, which you can access through their Web site at www.bsi.org.uk. Also, see Appendix 1 for relevant British Standards.

3 Baker, J R (1956) *The Use of English*, volume 8, number 1, page 36. 'English style in scientific papers.'

4 Beauchant, S (1999) 'Communicating by numbers', *Communicator*, volume 6, number 6, pages 14, 17.

5 Block, M and Brown, R (1999) 'Equations for nonmathematicians' *Conference Papers, ISTC Conference 99 Programme*, pages 22-24.

6 Swanson, E (1986) *Mathematics into Type: Copy Editing and Proofreading for Editorial Assistants and Authors*. American Mathematical Society, Providence, RI.

7 *Financial Times*. 18 June 1999, page 1. 'Misplaced comma costs Lockheed $70m'.

Further reading

Bly, Robert and Blake, Gary (1982) *Technical Writing Structure, Standards and Style*. McGraw-Hill, New York.

Bryson, Bill (1997) *Dictionary of Troublesome Words*. Penguin Books, London. (A witty, readable discussion of regular misuse of the English language.)

Butcher, Judith (1992) Copyediting: *The Cambridge Handbook for Editors, Authors and Publishers, third edition*. Cambridge University Press, Cambridge.

The Economist Style Guide (1998) Profile Books, London. (A handy source for nontechnical, mainly British English usage, including a word of caution to editors on page 4.)

Fowler, Henry W, (edited by Burchfield, Robert W) (1996) *The New Fowler's Modern English Usage, third edition.* Oxford University Press, Oxford. (Much quoted and regularly updated judgements on the use of English.)

Jones, Dan (1998) *Technical Writing Style.* Chapter 12, 'Editing for Style'. Allyn and Bacon, Needham Heights, Massachusetts.

Kemnitz, Charles F, editor (1994) *Technical Editing: Basic Theory and Practice.* Society for Technical Communication, Arlington, Virginia.

Rude, Carolyn D (1998) *Technical Editing, second edition.* Allyn and Bacon, Needham Heights, Massachusetts.

Samson, Donald C (1993) *Editing Technical Writing,* Oxford University Press, Inc., New York.

Siegal, Allan M and Connolly, William G (1999) *The New York Times Manual of Style and Usage.* Times Books, Random House, Inc., New York. (A useful source for nontechnical US English usage.)

Sun Technical Publications (1996) *A Style Guide for the Computer Industry.* Chapter 2, 'Working with an editor'. Prentice Hall, Upper Saddle River, New Jersey.

Tarutz, Judith A (1992) *Technical Editing: The Practical Guide for Editors and Writers.* Perseus Books, HarperCollins Publishers, New York.

Weiss, Edmond H (1991) *How to Write Usable User Documentation, second edition.* Chapter 10, 'Editing: revising for readability and clarity'. Oryx Press, Phoenix, Arizona.

3
The design of an open learning course

Kate Cooper, MISTC, *an iconoclastic and eclectic consultant, advises blue-chip clients how to improve communication. She left Managed Learning in 1993 to develop her consulting work further, and now also designs and delivers programmes on innovation with the University of Warwick, for managers across the world.*

A team of us designed, developed and delivered a unique, paper-based opening learning course *Writing Scientific English* for a specific client. It was created in the early 1990s, to enable mid-ranking scientists in a global research company to generate good documents faster. It took over a year in gestation and testing. Now, nearly a decade later, it is still being thumb-marked, pored over, photocopied, argued over ... and scientists within the organisation are writing documents the company needs, quickly and well.

This chapter tells the story of what we made from a design perspective, *how* its design was effective in shaping the learning of its users, and what lessons there are in this story for the technical communicator.

Design considerations

'Open' learners

As far as open learning was concerned, the course had to be just that – open. Our target learners had to be able to approach it and use it in any way they needed. Each one would start with different needs. Some would need one part and not another, or perhaps one part only at a certain time in their careers. Moreover, each person would use the course differently, owing to their different ways of learning. Hence the design of the course had to encourage both flexibility of access and navigation through it, to meet all these diverse learning needs. Every one of the components within it had to be easily and equally accessible.

Coming as we did from Managed Learning, an established training consultancy with various links into universities, we were also aware that the more adults know about how they learn, and how they acquire new knowledge, skills and intelligence, the better their learning will be. Whatever design we came up with, the design itself had to be made explicit, so once we'd completed the course, we wrote an introduction, describing some of the issues I'm discussing here.

The problems of linearity

Given that learners would use different parts of the course at different times, any thought of *linearity* had to be discarded. Having a beginning, a middle and an end just wasn't feasible.

But our target audience were highly trained scientists, many with PhDs, all highly literate, their very professionalism honed within the scientific tradition of rationality and logic; that is, their minds were steeped in and sometimes constrained within linear ways of thinking.

Presenting them with something lacking in overt logic, or expecting them to feel at ease with apparent ill-thought out muddle or mess, might well have turned them off. We had to engage them, not annoy them, or frustrate them before they started.

Creating networked information

What we needed, of course, was for the information to be linked, each component networked with all the others.

Remember, though, this was the early 90s, when hypertext and the web, or notions of mass customisation of an information network were merely expensive glints in Silicon Valley eyes, or glitch-prone efforts from top military minds. Though we couldn't access the kind of intranet technology you can now, luckily Adobe and Quark had by this time, produced flexible, hugely sophisticated DTP packages, and word processing was nearly as good as it was going to get. This earlier software technology gave us the tools, literally at our fingertips, to produce what was, in essence, a paper-based hypertext system.

Design concept 1 – a journey

The need to network the information led to two basic concepts for the design, both of which we calculated would appeal to every single one of our learners. The first concept was to exploit the metaphor of a journey for a set of navigation tools. Stories and place locations are as hard-wired in our minds as, too, I suspect is the notion of a journey. The basic 'journey' metaphor allowed us all sorts of conceptual tools for the reader – maps with differing degrees of detail, routes through the course, different routes to get to the same place, fast-tracks for those in a hurry and so on.

This metaphor also allowed us to enhance another feature of the course. The issue that always needs to be addressed with open or distance learning is that, however good the materials are, this delivery system for learning can generate a lonely, frustrating and difficult learning environment, so dropout rates can be high. All this can be effectively mitigated by the right support, providing motivation and focus for the learners.

We felt the support of line managers was key anyway; this fortuitously provided us with an opportunity to integrate the acquisition of the learner's new skills directly into the business. Each learner was assigned a mentor, nearly always their line manager. At a pre-course introductory workshop, the learner and mentor together decided the learner's 'route' through the course, with, crucially, the learning 'milestones' determined by the production of successful documents demanded by genuine business need. This approach inevitably had the added benefit of more senior managers reflecting on their own writing skills, and how they could develop them.

Design concept 2 – the zoom lens

The second design concept we created was what we termed in the course as the 'zoom-lens effect'. We're all familiar with maps of differing scales – an A-Z street map, an OS map for hiking, the road atlas to plan a holiday trip, a map of Europe, the Americas, the world. Though there is some evidence that people relate to them with differing degrees of effectiveness, we knew that simple 'zoom' effect would work with this audience. Although they might not have been good map-readers (though I'd have guessed they were!) they were all trained users of microscopes. As scientists then, they were all-familiar with seeing the same things at different levels of magnification, and could recognise that different levels of magnification can provide radically different appearing images. Hence we were sure, before we began, that they'd be able to use our *dooby*.

The *dooby* was this icon-like symbol for the whole course:

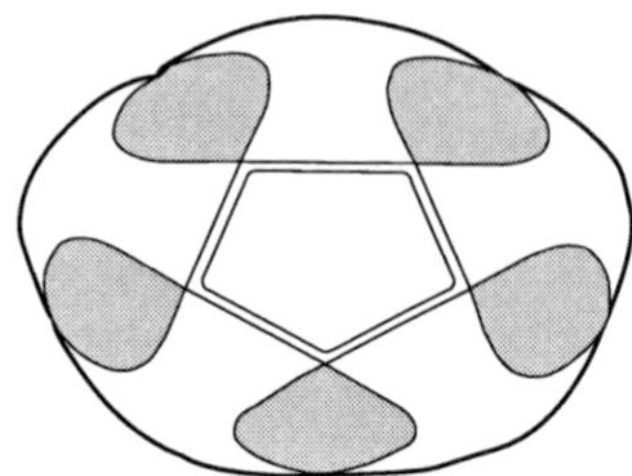

No one could think of what to call it officially. So our unofficial term for it just stuck. Dooby it became!

The dooby: the learner's map

The dooby was our base-line for the course 'map'. Just as all of us easily become familiar with certain map shapes, like the British Isles, say, or the map of the London Underground, so we estimated (correctly!) that our learners would become familiar with the dooby shape, and with the different versions of it, and hence where they were on their learning 'map'.

Their familiarity with it would also allow us to cut out detail not directly relevant to the precise place they were within the course, yet the overall shape of the dooby-map would let them know where they were within the context of the whole. Indeed, we sometimes superimposed a '*You are here*' sign on the dooby for this precise purpose.

We provided several versions of the dooby, each called a *Course Map*, each giving different levels of detail. On one map, learners had all the names of the modules, and on another all the units, too (as well as *Guidelines* which were templates for specific, tightly prescribed documents they had to write). Yet another Map had the module names, the units and a very brief summary of what was in each unit. We also made a version with 'empty' bullet points, and recommended that each learner made an A3 version of it, stuck it on the wall next to their workplace, and marked their progress through it, perhaps by filling in, or marking the bullet point in some way.

The choice of order within each list of units on the Course Map had nothing to do with the concepts within them; that is, the learners could, and did, approach them in any order. We did give some logic to the ordering of the list (as far as possible, they went from the general to the more specific) as we felt something completely arbitrary would annoy this particular readership! *But, as far as learning was concerned, each unit was conceptually independent.*

On the first page of each of the five modules, and on each unit (a sub-division of a module), there sat a highly simplified version of the dooby, with the '*You are here*' sign into it:

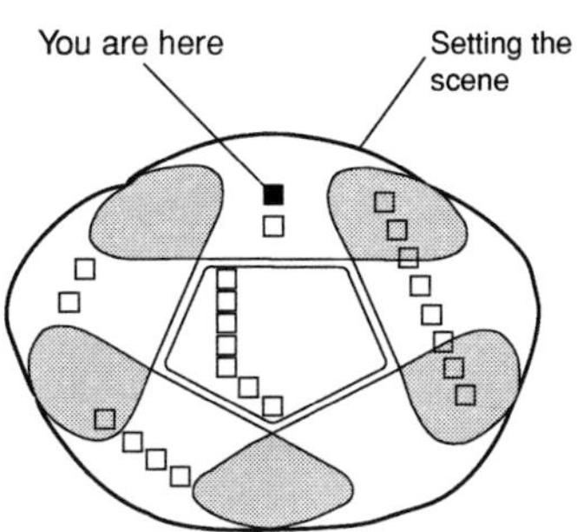

Learners would know, at a glance, where they were within the whole, and relate that to what they'd already studied and what they were planning to do next.

Circularity, linearity and repetition

However, despite our best efforts, we couldn't get away without using some linearity on the course. After all, page had to follow page. But that didn't mean learners had to start at page 1 and work their way through to the end. Pagination was *unit*-based, rather than module or course based. We reckoned, again correctly, that people would know where they were from headers and, crucially, from the dooby.

The dooby was deliberately a circular-like shape, with curves rather than lines, and with a swirling effect rather than any particular direction. Again, this forced the learners to approach the course differently from how they'd usually use a text, an approach without a writer-determined route or direction through it.

We were also keenly aware that each learner would approach the materials differently. Hence we provided *the same information in a variety of forms.* For example, we not only had the different versions of the dooby, giving different levels of detail, but also a Fast Track that contained almost word-for-word versions of the Unit Summaries. We hadn't a clue which versions of what were best, and still haven't! So we left it up to each learner to use whatever they found most useful, along with encouragement to experiment, and permission to discard elements that weren't helpful or appropriate for them at any particular time.

The 'look' of the course

We also had to make choices about typefaces and sizes, icons and how to indicate different types of information on the page. We had many discussions around this, but the information designer in the team kept asking herself two questions: First, will the text be readable and understandable? Secondly, will the pages look clean and accessible?

Typeface: Novarese

Lots of studies have shown that serif faces are easier to read than san serif ones. This may be because they are more familiar to us or, more probably, because the 'twiddly' bits, the patterns on the letters, give the visual cortex more information with which to distinguish each letter. However, san serif faces have a clean look, and are associated with modernity.

We actually chose Novarese, which is a serif face that has a large x-height. The strokes of the letters have a thick-and-thin element, which also eases pattern recognition. But Novarese, unlike the ubiquitous Times or familiar Garamond, doesn't have accentuated variation in the width of the strokes or the serifs. So it retains its visibility, and therefore its readability, yet looks clean on the page.

The designer also made the text a good size, and gave it lots of leading (actually 11 point on 14 point) – again, to draw the learner in, rather than be daunting.

Another attraction of using this typeface is that its *italics* version has an archaic look and is, in fact, quite difficult to read. We used Novarese Italics for suggested solutions and rewrites because its dense look and unusual serifs discouraged easy access. Its comparative unintelligibility meant that we could put such 'answers' to the side of the main text, knowing that the reader actually had to make a deliberate decision to get into them, and read them. Or they could, if they so choose, easily cover them with a bit of card or a bookmark. This placing of the 'answers' meant users didn't have to scrabble to the back of the text, or turn the course upside down, or whatever, in order to compare their ways of doing the tasks with the suggestions or recommendations of the course writer.

Amongst ourselves, we had discussed what approach to use at some length, and what the options were. Finally, we came to the conclusion that open learners should shape their own learning. The traditional approach of a learner doing a task, then assessing their 'answer' in relation to that of a specialist or expert, wasn't the only way of learning. Also, in the field of making language, compared to arithmetic or some of the sciences, there is no 'right' or absolute answer. Asking learners to tick or cross their suggested answer against something considered *The Answer*, was inevitably inappropriate. What we wanted to achieve was to develop their intelligence in criticising and learning from their own work, as well as from ours.

All this led us to decide that, if someone wanted to read the 'answer' before starting the task, they should be able to do so easily without any thought that such an approach was somehow cheating. (I was quite vehement in putting this argument forward at the design stage, for I had discovered while studying for an Open University degree, that reading the solution or discussion, before or during working on the set task, had often enhanced my learning, or made it much quicker.)

Finding a way to put our 'answers', whether discussion or a suggested draft text, conveniently very close to the task to which it related was important. Having Novarese Italic was excellent; the very detail of it would be easily ignored because of the difficulty people would inevitably have in perceiving words in this face, yet of course, the look of it naturally fitted well with the body text.

Typeface: Univers

We also needed, of course, to distinguish between different kinds of text. We chose to put larger heads and subheads in Univers Bold, so learners could find them easily against the background of the body text. Extracts from materials and sample texts appeared in Univers Condensed.

There were several reasons why Univers was chosen. It's a handsome, clean looking, san serif typeface, but not as ubiquitous as Arial and Helvetica are. It is also quite different from Novarese, yet the two faces complement, rather than clash with each other. Indeed, it looked far better with Novarese than did Gill Sans, one of the others we considered. As with Gill Sans, though, Univers is a member of a big family, so there were several variations of it we could use, including Univers Black, a splendid typeface for overheads and slides.

And, as luck would have it, when our client merged with another company a couple of years later, a new corporate identity was created, and Univers was one of the two typefaces chosen (the other being Bembo).

Typeface: Garamond

We wanted to make the Course Maps distinctive at a glance. So they, along with the frontispiece, were in Garamond. It is a richly rounded, welcoming typeface, especially in its higher point sizes. This enhanced the roundedness of the dooby.

The text, when there was some on a version of the Map, was in Garamond Condensed. This is a neat, small face with some of the qualities of its bigger family members, but allowing quite a lot of text in a small space and having excellent readability.

An added bonus was that Garamond was our own corporate typeface, though I suspect any branding effect was, if anything, minimal!

Using white space

I have long argued that good design is about *something and nothing*. And that the *nothing* (that is, the space on the page) is as important, if not more important, than the *something*.

Moreover, as I'm sure many of you are aware, research has shown that human eyes most comfortably scan an area 90-110mm wide. So we chose to make the main column of text 100mm wide, allowing lots of surrounding white space, and ensuring there were no big, off-putting blocks of text. This margin, too, gave us the space in which to put the 'answers' to tasks, and imposed a discipline on us that our 'answers' were short and pertinent!

Icons

When we were beginning to think about the design of the course, we imagined we would use several, maybe lots of icons, to delineate different kinds of text. We found, however, that a minimalist approach worked well. The typefaces indicated, at a glance, what was body-text, 'answers' to tasks, quotations and extracts, heads and subheads. We did feel, though, that learners would appreciate seeing where tasks were by simply scanning their eyes through the pages. So we did end up with one icon: to indicate a task:

The logistics of it all

The team

No one person has all the skills, experience and knowledge needed to devise, develop and implement such a course. A team was essential. On this team, too, we had crucially valuable contributions from trainers from the client company. Their role was to oversee the project from the client perspective, and to provide us with access to key specialist informants within the business.

On our side there was myself as the course designer. I also led the project from concept to delivery. Supporting me was Jane Cuthbert, then Account Manager (now Training & Development Manager with ICI Paints) who not only dealt with the day-to-day stuff, and kept track on all the minutiae such a project brings, but also somehow enabled every member of the team to translate the initial vision into tangible reality. On the team, too, we had Keith Richards, the Director of Studies at Aston's Language Studies Unit. He is not only an expert in linguistics and writing styles, but is also an expert on distance learning, leading a team at Aston writing such materials for their distance learning MSc students world-wide. Aly Robinette was our information designer; she took scribbles on the backs of envelopes, discussions and numerous meetings in her stride to produce the final design. Hardly mentioned in this chapter, but obviously of enormous importance to what we created for the client, was our writer Pieter Nelson; he had the hard, painstaking and unenviable task of matching our concepts and ideas with the needs of the client. A stunning achievement!

And the lessons for the technical communicator ...

There are several aspects of this story that have lessons for the technical communicator. First, and perhaps of most importance, is that *design for product* should be integrated into all the processes of creation or manufacture. The impact that design has on the final product is profound, and can (and should!) shape it from the concept stage right through to the final delivery.

Secondly, no one person is likely to have all the skills to take such a product from concept to completion. Having a team of people is important, and as important is investing in the development of that team. My Account Manager, in this instance, was invaluable; she not only kept us all to delivering what we'd promised on budget and on time, she also provided the social 'lubricant' if you like, that kept us all engaged and productive over its long gestation.

Thirdly, the look of little things matter, and can matter greatly. Like all primates, we're highly visual animals. So what something *looks* like affects the way we perceive it, and what information we get from it. We're used to getting a great deal of information at a glance (think how easily you recognise faces), so the design must have those detailed aspects that enable recognition and information transfer with effortless speed by the user. Things like typeface, column width, and easily identified text types really do matter in this regard.

Fourthly, you have to test ideas in the real world. We repeatedly went back to the client for information, for access to particular people or particular types of documents. We ran workshops for readers of documents, for secretaries who drew them up, for people who submitted multi-authored reports for outside organisations – we never stopped asking questions. And we built into the course easy ways of supplementing what we'd made. The client had a course that was also a framework that could be added to easily, as outside needs dictated.

Lastly, someone needs to have a vision of what the final product will look like, and what it will do for the client. And those involved have to trust the person with that vision, for there will always be periods of doubt and misgivings amongst the team. Holding that vision in mind is crucial. With this project, I was sometimes the only person who had this vision, certainly the only one who had it at the beginning – luckily for my sanity, I only discovered once we'd completed it... somewhat frighteningly, the client admitted to me that she'd had no idea what the final product would be like on commission, but had trusted that *I* had known!

What next?

I hesitate before making any prediction about the future, knowing full well that the only sure-fire prediction is that things will change. We also seem to overestimate what technological change will do in the short term, and underestimate what it'll do in the longer term.

With that caveat, I'll say this. There are fundamental design concepts that are well nigh universal and long-lasting, because they 'fit' our human perceptual system. Our human minds respond to them. Others prefer some shapes rather than others. We always like the touch, the smell and feel of natural materials – and the media we choose to access information from the world and each other will have to allow these qualities.

I like paper. I like books. I like browsing. I like the lack of time order in paper-based books and ephemera. I like the physicality of words and shapes. I like being able to mark a text. And I like having these kinds of things around me and in my living space. Will our children, and our children's children? I'll say this: whatever they'll be touching, using, feeling, they'll like the design of things to reflect a world that nurtures them, a design that 'fits' their age-old human mind.

4
Cross-cultural communication

Sandra Harrison, MISTC *is Head of Information Design at Coventry University and teaches technical communication on the Communication Authoring and Design degree course. She came to higher education from industry, where she worked in software documentation and was involved in the raising of documentation standards, including house style development, indexing, and user testing. Her interests include the development of student writing and email language.*

'In North Greenland distances are measured in sinik, in "sleeps", the number of nights that a journey requires. It's not a fixed distance. Depending on the weather and the time of year, the number of sinik can vary. Sinik...describes the union of space and motion and time that is taken for granted by the Inuit... The European measurement of distance, the standard metre in Paris, is...a concept for reshapers, for those whose primary view of the world is that it must be transformed. Engineers, military strategists, prophets. And mapmakers.'

Peter Høeg, Miss Smilla's Feeling for Snow[1]

You may be one of the many technical, professional and business writers today who are writing documents in English for an international audience. Your texts will be written and read in English, your British readers understand them, so is there a problem? Will other English speaking audiences understand what you have written? Will your words and sentences have the same meaning for English speakers from other cultures?

If your words have a different meaning or create different associations in another culture, your reader may be puzzled or amused. But the effects could be much more serious. If your product documentation causes offence you could alienate a major international client. If a mechanic misinterprets your instructions in an aircraft maintenance manual the result could be an air disaster. So how can you be sure that this will not happen?

When we write, we write from within a culture, and we naturally express that culture in our writing. Similarly, when we read, we draw on our own culture to help us interpret the text. So first we need to understand cultural difference, and then we need to consider how this might affect our writing, and what we can do to write appropriately for readers from other cultures.

Understanding cultural differences

Our culture is expressed in very many ways. Some of these are external objective factors, ranging from small details to whole systems, for example:

- Formats for numbers, dates and addresses
- The shape of electrical plugs and sockets
- Units of measurement (miles/kilometres, pounds/kilogrammes, degrees Fahrenheit/Centigrade, and so on)
- Political, religious, and educational systems.

Other factors are internal and subjective, and affect our beliefs and behaviour, for example the attitude towards authority in the reader's culture: Can it be questioned? Do employees participate in decision-making? Should workers act on their own initiative?[2]

These internal factors can cause unexpected problems in communication. This is illustrated in an account of why Australians and Americans sometimes misinterpret each others' indications of friendship. For Australians a sign that you like them is that you are prepared to argue with them, while for Americans a sign that you like them is that you agree with them. 'While the American is seeking a topic to chat about, the Australian is seeking a partner to spar with. Thus the American finds the Australian intrusive, and the Australian finds the American boring.'[3]

The quotation from Peter Høeg's novel at the start of this chapter illustrates cultural difference operating on these two levels. On the surface level there is a difference in the unit of measurement, sinik or metres, but on the deeper level there is a difference in the perception of how this measurement relates to the world about us.

Internal cultural factors will affect our whole approach to life, and this in turn will affect the way we interpret text and the kinds of documents that we relate to most easily. To illustrate this, Tom Warren contrasts two cultures:

- In the first culture, the emphasis is on individuality. The focus is on personal achievement and self-reliance. Here learning improves one's control, time is a river, moving quickly, and there is a strong pressure to keep up
- In the second culture, the emphasis is on the group. Here learning helps the group, problems are solved through intermediaries, time is a pool, and there is no hurry.

An understanding of such cultural differences will affect the way in which writers present information, and even the types of documents that they write:

- In the first culture readers need quick reference guides, overviews and summaries
- In the second culture the overall picture must be built up slowly.[4]

Readers from different religious and political backgrounds will have differing reactions to many topics which perhaps the writer assumed to be neutral: food (for example, references to eating roast pork), clothes (references to wearing a fur coat or shorts), the desirability of free elections, and ideals of individual success or freedom of thought and action.

The following sections discuss aspects of our writing that we need to consider from a cultural perspective.

Use of language

Our first challenge is to use appropriate language. One regional dialect may not be understood by speakers of another, but most speakers of English are familiar with the vocabulary and syntax of Standard English. The written forms of the language are more widely understood than the spoken forms – native English speakers from different parts of the world and speakers for whom English is a second language will all be familiar with written Standard English. In general, people who speak different varieties of a language show more differences in their spoken language than in their written language and written language is subject to much greater control, first by teachers and then by editors and publishers. This suggests that we should not move too far in the direction of writing in the way that we speak. For an international audience we should use Standard English rather than a non-standard regional or social variety. We should also avoid syntax and vocabulary specific to spoken language (for example, colloquialisms and contractions) even if we are constructing a form of direct speech, such as questions and answers, to convey information.

However, within written language, we have limitless possibilities for our choice of vocabulary. To what extent should we exercise this freedom of choice when writing for other cultures? In British schools, children are expected to vary their language. Teachers discourage the use of the same term in the same or consecutive sentences, and ask young writers to find different expressions. As a result, adult writers also expect to avoid repetition in their prose. But it has often been suggested that unnecessary variation of terminology to describe the same item or convey the same concept can be a source of potential confusion for the reader. Should writers then consider restricting their use of language in some way?

Humour

Although there are jokes that transcend national boundaries, humour relies on knowledge that is shared between writers and readers. It depends upon an understanding of both the external and internal aspects of a culture, and also an idiomatic knowledge of the language, so that the appearance of an unexpected word or phrase can be appreciated.

Humour can arise from many factors. Three common sources are:

- Ambiguity arising from a word that has more than one meaning, as in the very different meanings of 'club', below:

  'Do you believe in clubs for young people?' 'Only when kindness fails.' W.C.Fields[5]

- Ambiguity arising from different grammatical constructions that take the same form:

  'The students are revolting', where the phrase 'are revolting' can be interpreted as a verb 'are revolting', or the verb 'are' plus the complement 'revolting'.[5]

- Variations of well-known sayings:

  'Eat, drink and be merry for tomorrow we diet.'[5]

All three of the above types of humour can be problematic to readers from a different culture. The first relies on readers being able to draw simultaneously upon two very different meanings of the same word; the second relies upon them being able to recognise two valid grammatical interpretations and the third relies upon them recognising the well-known saying and appreciating the element which has been changed.

If we know a culture well, we might choose to enliven factual information by the judicious use of humour, but when writing for readers of a different culture we should exercise extreme caution. What might be considered funny in our society could be misunderstood, even to the extent of being found highly offensive, in another culture.

Metaphor

Metaphor can often be used to good effect in expository text by describing a new or difficult concept in terms of a familiar one. However, we need to be sure that the 'familiar' concept really is familiar in the culture of our readers. When we describe what we see on our computer screen as a 'desktop', we are assuming that our readers are more familiar with a desktop than with a computer, and that they will use this knowledge to extend their knowledge of the computer. But if the readers have no knowledge of desks, wastebaskets, folders and files, this concept would hinder rather than help their understanding. Any use of metaphor must be grounded in the readers' culture, using concepts with which the readers are thoroughly familiar.

Tone

Even when writing within a well-known culture, writers must make decisions about tone – about how formal or friendly, how serious, authoritative, chatty or ironic they will be. Thus most writers would not use the same tone when writing for local government treasurers as they would for teenage computer games enthusiasts. But although a writer might have a good understanding of appropriate tone for an audience that belongs to the same culture, readers' expectations about tone are different in different cultures. For example, British undergraduates often find American undergraduate textbooks uncomfortably familiar in tone:

'To see how thinking about your readers' tasks can help you write effectively, consider a report Lorraine must write. Lorraine works for a steel mill that has decided to build a new blast furnace...'[6]

This is the first time that the character Lorraine is referred to in this textbook, yet the very first reference to her, 'consider a report Lorraine must write' brings in her name in a way that indicates to a British reader that she is a character that the reader already knows. Moreover, it uses only her first name. In some genres this would be unremarkable, for example in a novel or a textbook for a young child, but a British undergraduate audience would probably expect the case study to be more formal, introducing the character when she is first mentioned, and giving her last name as well as her first. So for a British audience one might write, 'Lorraine Smith works for a steel mill that has decided to build a new blast furnace. She has been asked to write a report...'.

In this case, the effect of the familiar tone is that British students attempting to use the textbook might believe that it is too simple for them and fail to take it seriously, even if the content is appropriate and relevant.

Style

Readers also have culturally determined expectations of appropriateness in style. In business and professional communication, the western preference is in general for directness and specificity, and writers are expected to avoid digression. This relates to the awareness in western cultures of the cost of time, of the need to carry out tasks efficiently with minimum of errors or requests for help, and also of the awareness of the need to comply with safety legislation, to clearly specify correct and safe procedures, to describe explicitly any dangers, and clearly state any warnings and cautions.

Most western instruction to professional writers encourages us to use plain language to make the meaning of the text clear and unambiguous, but this is not necessarily seen as an objective in other cultures. 'Some cultures hold ambiguity or obscurity to be marks of wisdom because they are associated with knowing how to manipulate language.'[7] The Japanese preference for ambiguity over directness is well known[8], and we need to be aware that the use of directness in a culture that values indirectness can be seen as intrusive and impolite.

We also expect professional communication to avoid exaggeration and to be factual. The following extract from a letter of reference received from Iran conflicts with our expectations by using a non-western style based on inflated terms and generalisations:

> 'Now that Afshin wants to continue his studies it is impossible here in Iran due to the political turmoil, unrest, war, religious persecutions and the discrimination against the religious minorities.

> Therefore he wants to continue his studies in a country like the United States where they draw tremendous power from "Faith and Freedom". So then any help done to him will be a service done to the reward and enrichment of the humanity at large...'[9]

This raises another question: when writing in English for non-native speakers to read in English, should we adopt the conventions of our readers' culture, or maintain our own? If we were writing a letter of reference for a British student who wished to study in Iran, ought we to follow the style used above?

Structure

We must also consider the structures of our documents. Should they address their reader and approach their topic directly or indirectly? What makes a convincing argument to our readers? What counts as relevant, and what kind of evidence is needed?

Smith[3] suggests that the Japanese favour a spiral structure, while the British prefer a linear one. The Japanese supply 'all conceivable facts and ideas', which may seem irrelevant to the British. The result is that Japanese explanation seems to the British reader to be long-winded and off the point, while British writing seems to the Japanese reader to be too abrupt and impolite.

What difference does it make if readers are faced with an unfamiliar structure? To examine this, Ulijn[10] carried out an experimental study in which Dutch and French writers prepared four versions of instructions for a coffee-making machine. There were two 'originals' – one written in Dutch by a Dutch author and one written in French by a French author. The French version was then translated into Dutch, and the Dutch version into French. Thus the Dutch users had two Dutch versions – the 'original' Dutch version, and the version translated from French into Dutch. Similarly the French users had two French versions – the 'original' French version, and the version translated from Dutch into French. In each case, the main difference between the 'original' and the translated version was perceived to be the document structure.

The Dutch users understood both versions, and were able to carry out tasks effectively using both versions. The French users also operated equally well with both French versions. However, in general, the Dutch users preferred the 'original' Dutch version, and the French users preferred the 'original' French version. Ulijn attributes this preference to the structure of the documents, that is, the French users preferred the French structure, and the Dutch users preferred the Dutch structure.

This was a small-scale study and the effect of structure on the users was not very great, but the result is borne out by other findings where the effect was highly significant. For example a sales letter from an American company written in Chinese and sent to several Chinese firms received no response from Chinese recipients until it was rewritten with a Chinese structure.[11] These and other studies demonstrate that the structure can be important in shaping the readers' attitudes to the text, even when it has no significant effect on comprehensibility.

Using graphics

Given these problems with language, we might be tempted to think that we can resolve them by the use of graphics. But these too are culturally dependent. We create images that reflect our culture, and we interpret images from within our culture. We must therefore be as careful with our use of graphics as we are with our use of language. (Clip art must be reviewed as thoroughly as purpose-made graphics in this respect.)

As with language, we might use images which are culturally specific at the surface level, for example an image of a piece of equipment with a British-type plug. This particular example is unlikely to cause misinterpretation or offence, but it can cause a feeling of distance in the non-British readers because it is clear that the image was not produced with them in mind.

Other images may cause problems at a deeper level. For example, different cultures have differences in what is acceptable when representing the human form. (Is it permissible to represent both males and females? What activities are permissible? How much of the body must be covered? Does the culture attach an unfavourable significance to the representation of the left hand? – and so on.) Some cultures do not permit the graphic representation of the human form at all. Ignoring these conventions can cause serious offence.

Some images may cause difficulties in understanding. Icons are a frequent source of problems. While many well-known and internationally-used icons exist, a new icon which has not been thoroughly tested may be interpreted differently among different users.

William Horton provides a very useful introduction to the use of graphics in international documents.[12] He provides illustrated examples of the above points, and alerts us to other considerations such as the user's direction of reading (which can result in a series of images being read in a sequence not intended by the author), and the use of colour (colours have different associations in different cultures).

Possible solutions – internationalisation or localisation

There are two radically different approaches to the problem of writing for audiences of different cultures. The first approach is that of internationalisation. This approach seeks to eliminate anything which is likely to cause problems – metaphor, humour, variation of expression, and so on. The second approach is that of localisation. This seeks to develop in those who are writing and preparing the text a thorough understanding of the needs of the audience so that culturally specific elements can be integrated into the text according to the needs of that audience.

Internationalisation

The goal of internationalisation is to remove all cultural context, leaving 'a core product'[13] that will be acceptable across the world. However, it is very difficult for writers to identify all references that are specific to their own culture and it is necessary to obtain assistance in this from representatives of the users' cultures. Hoft[13] has useful advice about carrying out an international user analysis, and about performing cultural edits on your documentation.

Controlled English

The most extreme form of internationalisation is to use a controlled version of English. Controlled versions of several major languages have been developed for the purposes of international communication, and there are several different controlled Englishes, developed by different organisations. One of the best known is Simplified English, developed under the auspices of AECMA (the European association of national aerospace societies) and AIA (the equivalent body in the US). Simplified English is used by most major airlines and aircraft manufacturers. Writers may use only those words specified in the Simplified English dictionary (approximately 950 words). Each word may be used with only one specified meaning and as one specified part of speech. There are also numerous rules controlling syntax and usage, for example the '-ing' form of the verb is not allowed.[14] There are several advantages of this approach:

- Simplified English is clear, and successfully eliminates most sources of ambiguity
- It eliminates some problem areas such as colloquialisms and humour. (It would completely eliminate the first two types of humour given in the humour section above.)
- Simplified English is easier for non-native speakers to learn. (English is highly irregular and difficult to learn because it takes its roots from many different languages.)
- It aids machine translation through its restricted vocabulary and simplified grammar.

On the other hand, the use of a controlled language brings a corresponding loss. Restricting the vocabulary and grammatical constructions means a loss of the richness of the English language; clarity is achieved at the expense of variety; a unified tone is gained at the expense of lively prose. Certainly, it can be argued that for some purposes, for example in technical manuals, this is an acceptable price to pay for the benefit of clarity and consistency. However, it must be remembered that a controlled language like Simplified English does not take into account the needs of readers from different cultures – simplified English is controlled in ways that are determined by authorities from the writer's culture.

Localisation

The alternative approach to the problem of writing for audiences of different cultures is localisation. Writers or other designated members of their organisations find out as much as possible about the readers' culture and adapt the text accordingly, providing adequate context for out-of-culture concepts, ensuring that metaphor and humour are culturally appropriate, changing the structure of the document, and even inserting extraneous material where this will make the text more acceptable to its readers. To do this thoroughly requires a considerable investment of time and money, but the benefits can also be great because the finished product is thoroughly adapted to the users' needs and situation.

Choosing your strategy

How we decide between these options depends on the purpose of our writing, the expected audience, and the number of target cultures:

- Internationalisation is desirable when the documents will be used by readers from a variety of different cultures, and where budgetary constraints limit the number of versions that could be produced
- Controlled English is a suitable choice for technical experts carrying out well-defined technical tasks, as in aircraft maintenance manuals where the subject matter is restricted and the readers, although they may come from a wide range of cultures and native languages, will nevertheless have a detailed understanding of aircraft maintenance
- Localisation is more suitable for writing aimed at an audience from one clearly defined culture. It is necessary when the attitude of the readers to the text is important, for example when the text is written for a

sensitive audience such as those involved in business or political negotiation.

A further option is to combine the techniques of internationalisation and localisation. Here one might begin by creating an international document with the culturally specific features eliminated as far as possible. Those culturally specific elements which remain (for example, currency or date format) would then be listed, so that they could easily be adapted for different cultures.[13]

In general, this would lead to localisation at the level of surface detail. But it also makes possible selected localisation at a deeper level. It would be possible to build up knowledge and relationships with specific cultures from among the global audience, and to use the knowledge of these cultures to carry out more in-depth localisation of the documentation for these particular groups. The steps in this process would be:

1. **Global document**

   Create a global document stripped of all culturally specific references.

2. **Surface localisation**

   Add a limited set of culturally specific references to the global document (for example, date and currency references).

3. **Selected in-depth localisation**

   Identify a significant cultural group among the global audience, and adapt the original global document to generate a localised document for this group.

Finally, whichever approach you select, it is important to involve representatives from your users' cultures. This can range from full consultation with many representatives at all stages of the project, to a reading of your final draft by one potential user. The level of involvement will depend on such factors as whether you have chosen internationalisation or localisation, the importance of the project, and whether you wish to build a continuing relationship with users from a specific culture.

Cultural sensitivity and consultation with users will help you to avoid the pitfalls of cross-cultural communication and build a positive relationship with your global audience.

Resources

Useful web sites include:

Nancy Hoft's web site (includes a very good selection of links to various aspects of cross-cultural communication):
http://www.world-ready.com/readlist.htm

A scholarly introduction to controlled languages, with a useful bibliography and links:
http://www.mri.mq.edu.au/ltg/slp803D/class/Altwarg/index.html

Translation web site:
http://www.xlation.com

The ISTC site has links to other translation and localisation sites:
http://www.istc.org.uk/

The STC site has a dissertation database that includes abstracts, articles and conference papers, many of which focus on cross-cultural communication. The database can be searched by keywords:
http://www.stc-va.org/

A substantial bibliography is provided in Deborah Andrews' book.[15] This divides resources into printed material, internet resources and organisations.

References

1 Høeg, Peter (1994) *Miss Smilla's Feeling for Snow*, Flamingo, London

2 Cook, Rae Gorin (1996) 'Enhancing the Participation of Foreign-Born Professionals in US Business and Technology', in Andrews, Deborah C (ed.) *International Dimensions of Technical Communication* Arlington Virginia, Society for Technical Communication

3 Smith, Larry E (1987) 'Structure and argument' in Smith, Larry E (ed.), *Discourse Across Cultures: Strategies in World Englishes*, Prentice Hall, Hemel Hempstead

4 Warren, Thomas L (1993) 'Issues in Internationalization of Technical Documentation: Quality Control', *Proceedings of the Conference on Quality of Technical Documentation*, University of Twente, Enschede, The Netherlands

5 Ross, Alison (1998) *The Language of Humour* London, Routledge

6 Anderson, Paul V (1995) *Technical Writing: a reader centered approach*, Third Edition, Orlando, Harcourt Brace Jovanovich

7 Dissanayake, Wimal and Nichter, Mimi, (1987) 'Native Sensibility and Literary Discourse' in Smith, Larry E (ed.), *Discourse Across Cultures: Strategies in World Englishes*, Hemel Hempstead, Prentice Hall

8 Kohl, John R, Barclay, Rebecca O, Pinelli, Thomas E, Keene, Michael L, and Kennedy, John M, (1993) 'The Impact of Language and Culture on Technical Communication in Japan', *Technical Communication*, February 1993, Volume 40, No. 1, pages 62-73

9 Mirshafiei, Mohsen (1994) 'Culture as an Element in Teaching Technical Writing', *Technical Communication*, May 1994, Volume 41, No. 2, pages 276-282

10 Ulijn, Jan M (1996) 'Translating the Culture of Technical Documents: Some Experimental Evidence', in Andrews, D. (ed.), *International Dimensions of Technical Communication*, Arlington VA, Society of Technical Communication

11 Ulijn, Jan M (1995) 'The Anglo-Germanic and Latin concepts of politeness and time in cross-atlantic business communication: from cultural misunderstanding to management success', in *Hermes, Journal of Linguistics* no. 15, 1995 pages 53-79

12 Horton, William (1993) 'The Almost Universal Language: Graphics for International Documents', *Technical Communication*, November 1993, Volume 40, No. 4, pages 682-693

13 Hoft, Nancy L (1995) *International Technical Communication* New York, John Wiley

14 Farrington, Gordon (1996) 'AECMA Simplified English: An Overview of the International Aerospace Maintenance Language' in CLAW 96 *Proceedings of the first international Workshop on Controlled Language Applications*, Leuven, Centre for Computational Linguistics, Katholieke Universiteit Leuven

15 Andrews, Deborah C (ed.) (1996) *International Dimensions of Technical Communication* Arlington Virginia, Society for Technical Communication

5
Choosing language for effective technical writing

John Kirkman, FISTC *was Director of the Communication Studies Unit at the University of Wales Institute of Science and Technology, Cardiff (now University of Wales, College of Cardiff) UK. Since 1983, he has worked full-time as a consultant on scientific and technical communication. He has consulted for more than 300 organisations in 20 countries.*

General policy for writing readably

Two main aspects of the way we compose texts cause difficulty for readers: one is the choice we make of individual words; the other is the way we blend those words into structures – into phrases, clauses and sentences. In general, to reduce the reader's difficulties to the minimum, our policy should be to use 'everyday' language wherever possible, and to ensure that the statements we construct are neither too long nor too complex for easy comprehension.

Choosing words

Our overriding objective in writing about science and technology must be to convey meaning accurately. Accordingly, in any debate about whether one word would be shorter, more 'everyday', or more readable than another, we must always first establish whether the words have near enough the same meaning for the given audience and context. For example, I do not believe *utilize* and *use* carry different meanings for most readers, so I would always advocate the choice of *use* in technical writing; I do not believe *deleterious* means more or less than *harmful* to most readers, so I would always advocate the use of *harmful* in technical writing; but there are many occasions on which to change *rectify* to *correct* would be to alter seriously the technical meaning of the statement. To make such a change automatically on grounds that *correct* would be shorter and plainer would be disastrous.

In most technical writing, use of a great many 'heavyweight' words is unavoidable. If we are writing about *austenitizing* or *electrophoresis,* we have no alternative to using those terms (with clear definitions, if necessary); but we should avoid *unnecessary* specialist terminology and *unnecessary* heavyweight words. There is no virtue in using *manifests* if *shows* expresses the same meaning adequately. Indeed, since use of many heavyweight words is unavoidable, there is distinct benefit to be *gained* from consciously looking for opportunities to write in the shortest, most familiar words available:

> **Don't write...**
>
> The ABC assay was successful in identifying the presence or absence of LPS in the separated fractions. The presence of LPS was manifested by a low antibody titre when LPS was incorporated in the diluting buffer, while at the same time exhibiting a high titre in the absence of the 'inhibitory' LPS.
>
> **Write...**
>
> The ABC assay showed whether or not there was LPS in the separated fractions. The presence of LPS was shown by a low antibody titre when the diluting buffer included LPS, and by a high titre when no 'inhibitory' LPS was included.

By consciously adopting this style, we reduce the 'weight' of what we give our audiences to read. You may think that to take account of the number of syllables we ask our readers to handle is to become far too fussy about linguistic choices. Of course, when we consider the 'heaviness' of individual words like *utilize* and *use*, it is clear that occasional use of *utilize* instead of *use* is not going to ruin the readability of a text. But it is the *cumulative* effect of attending to many small matters like this that makes writing readable rather than unreadable. I cannot suggest in this chapter just one major thing to do to transform your writing dramatically. I *can* offer you many small pieces of advice on tactics that *cumulatively* will produce readable writing.

So, pay attention to the weight and familiarity of the language you use for your given audience. Undoubtedly, as readers become accustomed to individual heavyweight words, those words seem easier to handle. The word *electricity* (five syllables) is a heavyweight word, if we judge it on number of syllables alone. But heaviness is derived not only from number of syllables but also from what we can call the 'information load' carried by a word – from the difficulty and familiarity of the concept it expresses.

Most of us are familiar with the basic idea expressed by the word *electricity* (although few of us could define it rigorously!), and since it is a word we meet fairly frequently in daily life, we are not too surprized when we meet it in a text. We identify it without conscious effort, and move on rapidly to see what words lie ahead in the rest of the statement. But when we meet a word that we are not accustomed to handling, we are obliged to halt and to examine it for a little longer than usual. We 'work out' what it means, and then move on. If a text contains many words that make us halt to work out what they mean, we gradually feel a sense of struggle.

Often, we recognize that the writer was obliged to introduce terms unfamiliar to us: part of the task of acquiring new information is the learning of new terms to express that information. But equally often, I feel that I am having to work unduly hard to grasp the meaning of a text. I feel that other words could have expressed the meaning equally well, and would have demanded less effort on my part. No doubt you have sometimes felt the same. Try to prevent your readers feeling that they are having to work unnecessarily hard to get at the meaning of your text:

Don't write…

This allows the disk drives to be allocated between two disk processors giving an advantageous throughput enhancement as well as permitting up to twelve disk drives to be connected.

Write…

This allows the disk drives to be allocated to two disk processors, increasing throughput usefully, and enabling up to twelve disk drives to be connected.

Slang and colloquialisms

Let me emphasize that I am in no way advocating loose, slangy, excessively casual diction. Language must be used with maximum accuracy in scientific and technical writing. Slang and colloquialisms are usually inappropriate for technical work, because they are imprecise. They rely on shared understanding of background and nuances.

Here are examples of writing that is too loosely colloquial for reliable technical communication, especially in documents that will be read by people for whom your language is a second or foreign language:

> ... we feel that company X *has the edge on experience* in designing precalciner systems ...

> ... there will be a *large civil engineering impact* on the price ...

> ... ink wheels *are good for* several months of normal printing ...

> ... the need for *a decent packaging material* becomes more and more necessary ...

> ... location A and location B should also *watch* the A:B ratios ...

Local jargon and abbreviations must be avoided too. Will your audience recognize without difficulty such expressions as:

> ... using a *potmeter* ... (potentiometer)

> ... place the *sw.* in the ON position ... (switch)

... ensure that the *wng lt* is lit ... (warning light)

Even an expression such as: '... can improve beam focus, particularly with doughnut or multi-spot beam propagation...' will cause difficulty for readers who are unsure whether a standard doughnut is solid (and filled with jam or jelly) or has a hole in the middle!

Writing for international audiences

Writing for international audiences calls for special control over vocabulary and phrasing. Readers for whom English is a second language may have difficulty in detecting the different nuances you wish to convey by using a range of verbs such as:

place
position
locate
install
fit — the plate on...
mount
fix
put
apply

Do you wish your readers to do different things when you write: *power up, turn on, connect up, plug in, apply power to* an electric appliance? Do you wish to convey the same meaning when you tell readers to '*observe* the regulations' and when you write '*observe* the change in temperature'? What do you mean when you write:

... carefully *lower* spindle into frame ...

... *lower* back pressure of all output ...

... the *lower* speed is used ...

... for *straight* mineral oils ...

... marks are in a *straight* line ...

... the adaptor should have the *straight* side up ...

Two other features of choice of vocabulary are vital for clear communication to international audiences: consistency within documents, and standardization of terminology throughout the documentation of your organization. Be careful not to write 'moves towards the bowl periphery' on one page, and 'moves towards the bowl wall' on the next. Don't write 'the mill has had considerable down-time' on one page and then 'the mill has a low service factor' on the next page. Avoid calling a fastener a *latch* on one page and a *lock* on the next. Use *one* term consistently in your documents; be careful not to use *a video tube, a CRT* and *a picture tube* in different manuals.

Phrasing

The way we combine words into phrases needs as much care as we give to our choice of words, for two main reasons:

- Unusual phrasing – phrasing that differs from the norms of day-to-day expression – seems stiffly artificial to readers, and contributes substantially to the feeling that technical documents are difficult to read
- Complex word groups can be difficult for readers to comprehend.

A major cause of difficulty in technical writing is excessive 'premodification' – the piling up of modifying words in front of a noun. For example:

... a balanced ring trip detector input ...

... a complex frequency error correction procedure ...

Such chunks of linked words are difficult to digest. The reader has to keep eyes and mind open while waiting to see the whole related group *before* being able to work out how the words are meant to modify one another. Also, the reader's struggle is made worse by the fact that stringing so many modifiers together in front of a noun is not normal practice in everyday non-technical speech or writing. I suspect that you would find it unusual and clumsy if someone spoke or wrote to you about:

... a decorated playing card holder lid ...

Do these words mean that the playing cards are decorated, or the lid? The normal way of expressing that idea more manageably and without ambiguity would have been:

... a decorated lid for a playing card holder ...

In normal speech and writing, when we have two or more modifiers to add to a main noun, we usually put one or two in front, and the remainder afterwards. We take the two modifying ideas, *a decorated* and *for a playing-card holder* and place one on each side of the main noun, *lid.* Following that normal practice, we should present the technical statements as:

... balanced input for the ring-trip detector ...

... a complex procedure for correcting frequency errors ...

The main justification for recommending this style in technical documents is that it is the normal way of arranging phrases in English, and therefore does not seem artificial or 'heavy going' to readers. But there is another important justification. When a noun is heavily pre-modified, it is not always easy to see whether the pre-modifying words are meant to relate to one another, or individually to the main noun. For example, would you expect a pamphlet entitled *Guide to Successful Rabbit Raising* to give you advice on how to raise successful rabbits, or on how to raise rabbits successfully? Would this statement from a technical report introduce a statement about rectangular production of holes, or about production of rectangular holes: 'This would leave the six 514 machines dedicated to rectangular hole production'?

The addition of a hyphen in *Rabbit-Raising* would improve the title of the pamphlet on rabbits; and the sentence about holes would be improved by the addition of a hyphen between *rectangular* and *hole;* but the addition of hyphens still leaves readers to absorb large chunks *before* arriving at the main nouns they have to decode. Rearrangement to *Guide to Successful Raising of Rabbits* and 'This would leave six 514 machines dedicated to production of rectangular holes' is much more comfortable as well as unambiguous.

Expert readers (expert both in the subject-matter and in the reading of English) develop the ability to span three or four words at a time as they read. They can therefore comfortably read and understand small clusters such as *landing-gear system, cool-start solenoid valve, magnetic fuel-level indicator.* But even expert readers find it difficult to process long strings of words such as: '... outer wing-tank diaphragm-actuated-type shut off valves ...', especially if punctuation signals are omitted. Less expert readers (less expert both in the subject-matter and in the handling of English), who use a slower, word-by-word processing strategy, have special difficulty with such strings of words. Since so many technical documents are now produced for international distribution, it is wise to use no more than two pre-modifiers wherever possible.

Clusters of nouns

In particular, it is wise to avoid confusion when you use a noun to modify another noun. In English, a pre-modifying noun, normally (not always, but normally) is equivalent in meaning to a post-modifying *of* ... phrase:

waste disposal = disposal of waste

water repellency = repellency of water

grass cutting = cutting of grass

For this reason, when we see signs outside a factory, 'CUSTOMER RECEPTION' seems normal (= reception of customers) but 'CUSTOMER PARKING' raises a smile (= parking of customers!). In general, it is more accurate and more readable if you use *of* ... constructions instead of excessive pre-modification of nouns:

Don't write ... enables traffic handling optimization ...

Write ... enables optimization of traffic handling ...

Or better ... enables optimum handling of traffic ...

Don't write ... influences measurement accuracy...

Write ... influences accuracy of measurement ...

Don't write ... is used for interrupt handling ...

Write ... is used to handle interrupts ...

Don't write ... methods for file updating ...

Write ... methods for updating files ...

Don't write ... is important in sub-station site selection...

Write ... is important in selection of sub-station sites.

Clusters of nouns can cause considerable difficulties, especially for translators. Would you be sure how to convert the following original expressions into another language?

Original ... system for automatic land vehicle monitoring

?= ... automatic monitoring of land vehicles ...

?= ... monitoring of automatic land vehicles ...

Original ... extracts data from the failed primary controller memory ...

?= ... from the failed memory of the primary controller ...

?= ... from the memory of the failed primary controller ...

Prepositional phrases

One reason why clusters of nouns cause difficulty for readers is that pre-modifying nouns often come between a preposition and its true 'complement'. A preposition normally introduces (governs) a noun, a noun phrase or a clause (such as a *wh-* or an *-ing* clause):

Examples:

With	difficulty
At	the door
From	what has been received previously
By	lifting the reactor lid

In modern linguistic terminology, the noun, phrase or clause is called the *complement* of the preposition. The normal convention in English is that the first noun after the preposition is the complement. For example, sentence 1 below would seem normal, sentence 2 would seem abnormal:

1. ... ambient light and temperature are major influences on *the attentiveness* of pupils in classrooms.
2. ... ambient light and temperature are major influences on *pupil classroom attentiveness.*

Unfortunately, in a well-intentioned effort to economize in the use of words, technical writers have got into the habit of packing nouns between the preposition and its complement, as in 2 above. This causes at least reader discomfort (or better, discomfort for readers!), and at worst, a breakdown in communication:

Don't write ... follow the steps for monitor removal ...

Write ... follow the steps for removing the monitor ...

Don't write ... insert a probe under the high voltage anode rubber shield ...

Write ... insert a probe under the rubber shield of the high-voltage anode ...

Don't write ... the development of systems for marine traffic control ...

Write ... development of systems for controlling marine traffic ...

Don't write ... this stage of the project will be followed by control equipment selection and purchase ...

Write ... this stage of the project will be followed by selection and purchase of control equipment.

Sentence structure

Manageability must be the uppermost thought in our minds as we build our words and phrases into complex statements – into sentences and paragraphs. Broadly, we create sentences in one of two ways:

- By linking phrases and clauses in 'main + subordinate' relationships (complex sentences)
- By stringing together elements of roughly equal importance in a linear chain (compound sentences).

Neither technique of sentence-building is intrinsically better than the other. Both are legitimate ways of creating statements in English, and both can be used with clarity, incisiveness and elegance. But both techniques can be used clumsily. A writer who indulges in too much subordination or 'embedding' of clauses creates complex statements that make overwhelming demands on our short-term memories. We need to break them into more manageable groups for sorting and storage. Equally, if a writer links too many elements in a single chain, we feel that our mental jaws cannot open wide enough to swallow the information that is being rammed in. Eventually we choke, and have to break the meal into smaller 'bites' that are easier to chew, swallow and digest.

So, as you write, take care not to make statements that are too long and/or too complex. Check the 'unloading rate' at which you pass information to your readers:

Don't write...

... For a write operation, the user system first performs the 3 byte command sequence, as in the read operation and next initiates (asynchronously, if the MDC is operating in buffered mode) an X bytes transfer into the Data Buffer (with X = value of 1 to N x 256; N being the number of sectors to be chained), and then indicates the end of transfer with a TERMINATE command.

Write...

... For a write operation, the system:

1 goes through the 3-byte command sequence as in the read operation;

2 initiates an X-bytes transfer into the data buffer;

3 indicates the end of transfer with a TERMINATE command.

> In the transfer, X is the value of 1 to N x 256; N is the number of sectors to be chained. If the MDC is operating in buffered mode, the system initiates the transfer asynchronously.

Or perhaps

... For a write operation, the user system:

a) starts with a 3-byte command sequence (as in the read operation);

b) transfers into the Data Buffer 256 bytes for each sector to be chained;

c) ends the transfer with a TERMINATE command.

If the MDC is operating in buffered mode, the transfer of bytes is made asynchronously.

Style for instructions

Instructions must be unambiguous and specific. These qualities are derived mainly from clear thinking by the writer about what to say and how to arrange it; but errors of sequence, jumbling together of instructions and commentary, failure to specify who is to do things, and failure to provide '... if not ...' information are all common faults in technical manuals. For example, the following text purported to give instructions. It is grammatical and written in short, manageable sentences, but it is poor instruction-writing:

> The receiver can now be tilted backwards, as shown in Figure A, until its flange rests on the two supports. The chain safeguards the unit and prevents it from tilting further than necessary.
>
> Before executing this step, make sure that the chain is undamaged and reliably fixed.

Above all, this example does not obey the cardinal rule of instruction-writing: that instructions must instruct. To say 'The receiver can now be tilted . . .' is to *describe.* If you mean: 'Tilt the receiver ...', say so with an imperative verb.

Most instructions should begin with the imperative form of a functional verb *(raise, lift, turn, remove ...)*; but sometimes, it may be valuable to begin your instruction with an indication of time, a condition, a specification of a tool, or an indication of place:

- A time phrase/clause:
  'When the package arrives at X, change ...'
- A condition:
  'If the temperature drops more than three degrees, raise ...'
- A specification of a tool or piece of equipment:
  'Using extractor tool XY12, remove ...'
- A siting phrase/clause:
  'At Board C, switch ...'

Take care to avoid vague terms like *select* (does that mean I have a choice?), *replace* (does that mean 'put back' or 'fit a new one'?) and *check* (does that mean just 'look at' something, or 'do' something if a given condition is not being met?). Make sure that your statements do not seem to leave an option open: 'The flammable solvents *should be stored* in the solvent bin .. .' and that they cannot be interpreted as simple statements of future action: 'Acrylonitrile levels *will be monitored...*'

In writing instructions, we must take maximum care to ensure that our writing is accurate, explicit, and imperative.

Cryptic, abbreviated writing

In many technical settings, we want to be as terse and economical as possible, to save space. This applies generally in technical writing, but especially in writing instructions, writing memoranda, and writing for a computer screen. In these circumstances, we are often tempted to write cryptically, omitting small words like *a* and *the.* Unfortunately, this style can leave readers uncertain about what was intended. The following examples from instructions usually make readers pause:

> '... voltage values are seen through small windows in panel. Switch ranges from 100 to 240 in six stages, and is positioned by turning ...'
>
> '... To avoid damage to the actuator stem, supply air pressure to diaphragm must not exceed 15 lbf/in^2 and must ...'

The best policy is always to write everything in full, especially if your documents are likely to have overseas readers. Rarely will you find that your texts become significantly longer.

Personal and impersonal writing

Much report-writing is dull and ambiguous because writers try to strait-jacket the whole report into a single style: they try to use third person, past tense, and passive voice throughout the text.

It is usually impossible to write a report wholly in the past tense. Though *most* of a report will naturally look back and focus on what *was* done, almost always there are occasions when the writer wants to talk about generally accepted truths ('and because the gross output from x *is* y,..') which necessitates a move to the present tense (and other tenses) occasionally.

Especially in a *Discussion* section of a report, variation of tense is essential. But be careful to sustain a comprehensible sequence of tenses throughout a given section of your discussion. Here is an extract from a discussion of a scheme for the future; the re-development plan has been approved and is being explained in this report, so the correct tense through most of this section should be *future:*

> It is intended that the main pedestrian circulation routes are basically unaltered by the scheme. Pedestrians coming from Hog Hill will travel across the existing High Street Bridge into Someplace, although a crossing is required adjacent to High Grove House. Pedestrians coming from the Wharf area westwards either cross the new route at St Mary Street, or pass under it via the extended footpath on the river bank. Pedestrians will be discouraged from using the new bridge, and for this reason footpaths of

only 1m are provided. In other areas, footpaths are to be 1.8m wherever possible. It is felt that in removing trunk route traffic from the existing bridge and St Mary Street, the scheme offers considerable opportunity to improve the pedestrian facilities in the Promenade area. The existing bridge will be restricted to pedestrian traffic only, and the portion of Bridge Street between St Mary Street, and the High Street Bridge will be restricted to essential traffic only. The bridge would be resurfaced to improve its waterproofing and appearance. The paved portion of the Promenade between the existing bridge and the new crossing would be taken up and relaid as grass.

Revised version

We propose to leave the main pedestrian routes unaltered. Pedestrians coming from Hog Hill will travel across the existing High Street Bridge into Someplace, although a crossing will be required next to High Grove House. Pedestrians moving westwards from the Wharf area will either cross the new road at St Mary Street or pass under it on the extended footpath along the river bank. Pedestrians will be discouraged from using the new bridge, and for this reason, footpaths of only 1m will be provided. In other areas, footpaths will be 1.8m wherever possible.

We think that by removing existing trunk-route traffic from the existing bridge, the scheme offers considerable opportunities for improving the footpaths in the Promenade area. The existing bridge will be restricted to pedestrians, and the part of Bridge Street between St Mary Street and High Street Bridge will be restricted to essential traffic. The bridge will be resurfaced to improve its waterproofing and appearance. The paved part of the Promenade between the existing bridge and the new bridge will be taken up and grass will be put in its place.

To write in the third person throughout a report almost always produces clumsy style and often introduces ambiguity.

For example, in the extract above 'It is intended that the main pedestrian circulation routes are basically unaltered ...' is a clumsy and ambiguous way of expressing 'We propose to leave the main pedestrian routes unaltered ...'. 'It is felt that ...' is an unclear substitution for 'We think...'. 'It is felt ...' leaves ambiguity about who is doing the feeling/thinking.

Of course, there are sections of a report in which it is inappropriate to insert unnecessarily the fact that you personally were involved. In a description of an experimental procedure, it is usually irrelevant – would even give the wrong emphasis – to keep saying *I* or *we* did things. To throw emphasis mainly on the action itself, without reference to the actor, we use a passive construction: 'Reactor A was connected to Reactor B and oxygen at 7 lbf/in^2 was passed ...'. But this is not to say that you should *never* specify when *you* did something. It would be entirely proper to write 'The vessel overflowed after 15 minutes, so we asked X Department to allow us to divert...'. Indeed, to write '... so X Department were asked to allow us to divert...' would be to write less accurately and less clearly.

To write effectively in reports, set your thoughts clearly on what you were asked to do, how you did it, the results you obtained, and what you conclude/recommend: then write as plainly as you can, using as far as possible the language you would choose if you had the chance to present your account in a face-to-face conversation.

Again, let me emphasise that I am not advocating loose, casual diction. I am saying that it is unnecessary – is often distorting – to strain for a special style for writing reports. In face-to-face conversation, you would be unlikely to use the formal, roundabout expressions in the (genuine) extracts below. If you think the revisions seem more comfortable and natural, *and* express adequately the writer's meaning, draw the obvious conclusion – a style similar to that of the revisions is the style you should strive for when you write reports:

Original:

... authorise the insurers to effect a direct payment to X ...

Revision:

... authorise the insurers to pay X directly ...

Original:
... It has been shown that paraquat toxicity in rats is enhanced by exposure to a 40% oxygen atmosphere ...

Revision:
...We have shown that a 40% oxygen atmosphere makes paraquat more toxic to rats ...

Original:
... the site has previously been subjected to industrial activity including iron making ...

Revision:
... the site has previously been used for industry, including iron-making ...

Plain and manageable writing

Throughout this chapter, I have advocated that you should adopt a policy of writing 'plainly and manageably'. I am frequently asked three questions about that policy:

- Will such a style be generally acceptable in the scientific and engineering communities?
- Will use of such a style diminish the credibility and esteem accorded to me by my peers?
- Is the extra effort involved in thinking and writing more clearly *really* repaid – does it make *that* much difference?

In order to reassure scientists and engineers that the style I advocate is acceptable, I have made several surveys among professional scientific and engineering groups. All the surveys have produced clear majorities in favour of plain and manageable style. Full results of those surveys are given in *Good Style*[1].

Anxiety about possible loss of credibility and esteem is natural. After all, if we are realistic, we must recognize that some writing (much writing?) aims as much at the promotion of its writer (or of the writer's employer) as at the spreading of new knowledge. Also, much writing (for example, requests for capital, or proposals for projects) must present a convincing case and change minds – must persuade and impress in the best sense of those words. It is important that the language used should help to create a sense of conviction and confidence.

In *Does Style Influence Your Credibility?*[2] , Ewa Bardell presented evidence to show that plain and manageable style brings *greater* credibility and esteem than a more formal, heavyweight style.

All the research evidence I have ever seen suggests that plain and manageable writing brings a real return for the effort involved. I hope you, too, will find that to be true.

References

1 Kirkman, John (1992) *Good Style: Writing for Science and Technology*, Spon, ISBN 0 419 17190 8

2 Bardell, Ewa 'Does Style Influence Credibility and Esteem?', *The Communicator of Scientific and Technical Information*, ISTC, No. 35, April 1978, pages 4–7

6

Creating a style guide

Colin Battson, FISTC *began his career as an Aircraft Antenna Design Engineer with Standard Telephones & Cables Ltd, following a student apprenticeship. Seven years later he embarked on a second career in technical writing, beginning as a contract author on-site with Marconi Space Division. In the 30 years or so since then, his work has embraced a wide range of disciplines and topics, whilst employed for the most part by Technical Publications Consultancy companies. He is presently Chief Author for Author Services Technical Ltd, in Letchworth, Herts.*

Overview

To help you navigate through this chapter, read this overview first. If you are familiar with Style Guides and merely want to refresh your memory on a particular aspect, it may save you time. The chapter comprises a series of numbered sections, as follows:

Section 1

The first section defines the term 'Style Guide'. Do *not* confuse with 'writing style'. A Style Guide defines features such as page size, page layout, fonts, numbering schemes, and so on, whereas writing style is all about the way the paragraphs, sentences, and the words within the sentences are structured. Refer to Chapter 5 for further information on Writing Styles.

Section 2

We move on to giving some of the reasons why a Style Guide is needed, and how a well-defined one is of help to all concerned with the documents to which it applies. Not just to the author but all involved with the document production (such as word processor operators, proofreaders, graphic designers, typesetters or desktop publishers and finally printers).

A good Style Guide is also of considerable benefit to the reader or end-user of the document. Clear and recognisable styles in heading weights, fonts, contents lists, numbering systems and so on make it much easier to use a publication. A well-designed Style Guide can make the difference between the document being read and being rejected as unreadable.

Section 3

There are very few documents published that are not controlled by some form of Style Guide, even though there may not necessarily be an associated document actually called Style Guide.

Just think of company brochures for example. They are produced to have a strong corporate identity to promote the company. Features such as the colour, size and position of the company logo and the font used for the company name are all-important. Advertising relies for its success on clearly recognised corporate style. This section examines some of the elements covered by a Style Guide, and some pitfalls to avoid.

Section 4

This section discusses document structures such as chapterisation, use of annexes or appendices and so on.

Section 5

Illustrations and other graphics. Why and when to use them; which types are most suitable for different applications.

Section 6

Style Guide aspects and techniques particular to on-line documentation.

Section 7

This final section gives some ideas for the important aspects of distribution and issue control of your Style Guide. It is useless to produce a fine, detailed and comprehensive Style Guide if it is not circulated to all who should be aware of it; similarly if it is updated without informing all copyholders of the changes.

Now read on!

Section 1 - What is a Style Guide?

General

A Style Guide could be described as the group of rules that define your information or document design. Although documents produced for diverse purposes are likely to require differences between individual styles, many of the essential elements will be there; only the detail will vary. As examples you may specify your User's Guides to be A5 format, Wiro bound and in colour, whereas the Workshop Manuals produced in accordance with your Style Guide are A4 format, assembled in 4-ring binders, and in black and white.

Creating a new Style Guide – the essentials

If you are intending to create a new Style Guide, first consider the range of document types to be covered by it. The greater the range of document types, the more comprehensive and exhaustive your Style Guide needs to become. It is not uncommon for company Style Guides to be published in (say) a 4-ring binder between one and two inches thick. Planning is crucial. It is important to establish at a very early stage the full scope of the Style Guide, so that design elements and structure allow for all document types to be produced.

Make a list of all the elements that will need defining for the information designer who will use your Style Guide. The following may be considered as merely a typical list for hard copy documents, presented in a hierarchical manner, with the key features sub-divided into further detail. Dependent on specific requirements, your list may differ considerably and may be shorter or longer.

- Page Size
  1. Aspect (portrait or landscape)
- Page Layout
  1. Margins
  2. Headers and footers
  3. Columns (single or multi-column format)
  4. White space
- Chapterisation
  1. Numbering
  2. Appendices and annexes
  3. Contents lists
  4. Glossaries and indexes
- Fonts
  1. Types
  2. Sizes
  3. Warnings and cautions
- Illustrations and graphics
- Vocabulary and spelling standards (for example, UK or US English)
- Abbreviations - rules of usage
- Issue control
  1. Amendment instructions
  2. Amendment record
- Software and version (word processing, DTP, graphics, etc.)
- Storage media, archiving, etc.

Where a Style Guide caters for a range of document types, it could be structured to include a general or global section, followed by separate sections for each type of publication. For example, if it provides style rules for technical manuals, User's Guides, publicity or product brochures, in-house reports, contracts and tenders, Quick Reference Leaflets, and so on, it is clear that there will be conflicting requirements in terms of page size, use of colour, paper stock, etc.

Section 2 - Why have a Style Guide?

General

A well-defined Style Guide is of help to all concerned with the documents to which it applies. Not just the author or information designer, but all involved with the document production. This may encompass word processor operators, proofreaders, graphic designers, typesetters or desktop publishers and even printers. A good Style Guide is also of considerable benefit to the reader or end-user of the document. Clear and recognisable styles in heading weights, fonts, contents lists, numbering systems and so on make it much easier to use a publication. In some cases the existence of a clear structure can make the difference between the document being read and not being read at all.

Benefits to the author

- If an author follows the applicable Style Guide rules when creating a document or publication, all the basic document design parameters are predefined, saving precious time spent in 'playing' with different layouts at the outset
- If the author rigorously adheres to the rules, and is producing just part of a larger document or publication, the parts can be joined together 'seamlessly' to produce the whole.
  Note: Using the style gallery available in the word processing software will ensure full compatibility of texts produced by different authors, provided that:
  (i) The authors all follow the styles defined in the Style Guide, and
  (ii) The styles have been set accordingly in the word processors used by all contributing authors
- By following the prescribed Style Guide, an author can be confident that the style of the document produced will comply with house rules or corporate identity requirements of the issuing authority.

Typical house rules

Publication style:

- Sizes permitted/recommended (for example, A4, A5, custom)
- Paper and covers specification
- Binding method (for example, ring binder, comb bound, Wiro bound, perfect bound, etc.)
- Page layout standards
- Numbering protocols (document, section, page, paragraph, figure, table, etc.)
- WARNING and CAUTION standards (fonts, capitalization, symbols, borders, positioning, etc.)
- Standard preliminary pages
- Contents lists, glossaries and indexes
- Amendment policy
- Appendices.

Publication structure and content:

- Definition of publication type (for example, Workshop Manual, User's Guide, System Manual, etc.)
- Standardization of chapter titles per publication type
- Definition of content of each chapter type
- Chapter dividers (material, colours, tabs – titles or alphanumeric)
- Positions of figures (in-text and/or at chapter end)
- Examples of page layouts and illustrations
- Samples of information (for example, technical data, fault-finding guide, parts list, etc.).

The principal benefits of producing all publications to a well-defined Style Guide are:

- Easier planning and costing of new publications
- All contributors work to defined and consistent standards; keeps costs down
- Input from multiple authors/illustrators without major editing
- Publications have a uniform appearance and 'feel'

- End-users can navigate more easily when structure and numbering is consistent
- Material from one publication can be readily cut and pasted into another
- Printers can identify layout errors before expensive printing has been done
- Control of amendments and revisions is built-in from the outset.

Section 3 - Some Style Guide elements

Page layouts

Before settling on a preferred page layout for a particular document type, consider the purpose of the document, the typical end-user of it, the type of information to be presented, and whether the published documents are to be single- or double-sided. If the latter, page layouts must be designed for odd and even (or right and left hand) pages. New chapters or sections normally start on a right hand (odd numbered) page.

Margins and white space

Every page has areas of white space; all documents should be designed with areas of white space to make them visually appealing. Margins are one aspect of white space. Another is inter-line spacing, known as *leading* (pronounced *ledding* – a printing term originating from the days when strips of type metal were inserted between lines of text in typesetting). This is covered in more detail later in this section.

All pages have four margins, (see Figure 6-1):

- Top – also called head
- **I**nner – also called **L**eft side or binding margin in single-sided publications or gutter in double-sided publications
- **O**uter – the side margin away from the binding edge (always **R**ight in single-sided documents)
- Bottom.

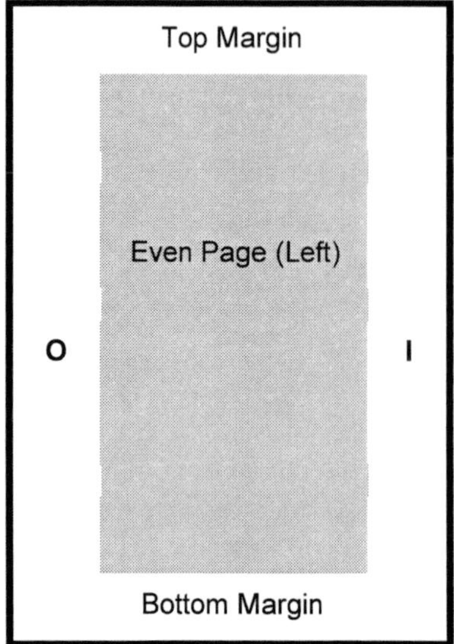

Double-Sided Publication Pages

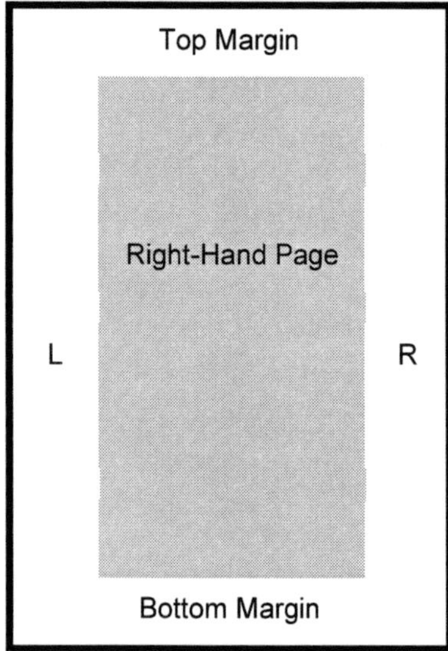

Single-Sided Page

Figure 6-1 Margins for double and single-sided pages

The Inner or Binding margin (always the Left margin for single-sided pages) must be of sufficient width so that printed material, text and/or graphics, is not lost, obscured or damaged in the binding process.

The Outer margin (Right margin for single-sided) need not be so wide, unless it is assigned for a particular purpose such as making notes or inserting marginal notes or graphics in the document itself.

Top and Bottom margins may contain headers and footers respectively (see later in this section).

Text justification

If all complete lines of type on the page are exactly the same length, and extend to both left and right margins, the text has been *justified* – implemented by adjusting proportional spacing between words. This gives the document a clean right-hand edge which does look neater, and is particularly useful in multiple column page layouts – see later. This paragraph is an example of justified text.

However, tests have established that fully-justified text is somewhat less easy to read, so improved appearance has to be weighed against readability. Fully-justified text is often used in multi-column layouts, where the right-hand edge of the text helps emphasise the column line.

If text is not right-justified, the style is referred to as 'ragged right edge'. This style does lend itself to easier revision, and – in a survey carried out in the USA – was favoured by the majority.

Leading (inter-line spacing)

Text is easier to read when line spacing is 1.5 times the character height. Spacing between paragraphs may be increased sufficiently above that, to be observed as such without conscious effort. Additional space between paragraphs is also an aid to faster and easier reading, as it gives the reader a natural 'pause for breath' point.

The above paragraph is repeated below with reduced line spacing, but using the same font and font size, to demonstrate the reduced ease of reading caused as a consequence.

Text is easier to read when line spacing is 1.5 times the character height. Spacing between paragraphs is increased sufficiently above that, to be observed as such without conscious effort. This additional space between paragraphs is also an aid to faster and easier reading, as it gives the reader a natural 'pause for breath' point.

Headings

Headings are used to label sections of text in a document, and by careful selection of a range of heading styles or 'weights', the author can use recognisably different heading styles to identify the relative importance of the subsequent text, thereby creating a structure to the document. The relative importance of individual styles should be designed to be self-evident by judicious choice of fonts and other text attributes.

Headings also serve to break the document into manageable chunks for the reader, again contributing towards making the document easier to read. There is nothing more daunting than turning a page to be confronted with a double-page 'spread' of solid text without the welcome interruption of a heading or two and paragraph spaces as natural breaks or pause points.

In a typical word processing software package, the properties of the various heading 'weights' (degrees of importance) are stored and indexed by the heading name, which can simply be listed as Heading 1, Heading 2, and so on. If this style gallery, as it is known, is promulgated along with the hard copy Style Guide to everyone involved in document preparation and production, headings will be of consistent style.

Following is just one typical example of heading weights for a range of five heading styles:

HEADING 1 – CENTRED, UPPERCASE AND BOLD TYPE

HEADING 2 - FLUSH LEFT, UPPER CASE AND BOLD TYPE

Heading 3 - Flush Left, Initial Caps. And Bold Type

<u>Heading 4 - Flush Left, Initial Caps. And Normal Type, Underlined</u>

<u>Heading 5: Flush Left, Initial Caps., Normal Type Underlined; In Line With Paragraph Text:</u> Xxxxxx xxx xxxxxxx xxxxx xx xxxxx xx xxxxxxxx.

Page grids

Each page can be envisaged as a grid structure, in which columns and rows are used to define and organise the text and graphics, including headings, headers and footers that make up the complete page. Different grids are appropriate for document types targeted at different audiences. For example, a single-column text grid may be used for technical reports or technical manuals, whereas a two-column variant may lend itself to an Operator's Handbook or Product User's Guide.

A three- or even four-column grid may be best for a corporate newsletter. The 'broadsheet' newspapers use as many as eight columns in their page grid.

This section of text has been produced in a three-column layout, illustrating the quite different appearance this part of the document has, compared to the single-column layout of the rest of this section.

The text in the first column is unjustified (ragged right edge), whereas the second and the third columns contain right-justified text.

Multi-column layout has the obvious disadvantage that line lengths are shorter, hence readability is reduced unless font size is also reduced considerably. That can also cause readability problems in itself, so is not to be recommended unless absolutely unavoidable.

Observe from the example of three-column layout on the previous page that justified text defines the columns better, but can give a somewhat awkward appearance when inter-word spacing becomes expanded as a consequence. All these factors must be taken into account when designing page layout standards for your Style Guide.

Numbering schemes

Almost all publications require one or more numbering schemes, in order to aid reader navigation and to simplify cross-references. It is important that the Style Guide makes allowance for the full range of numbering systems that are likely to be needed for the documents produced to its standards.
Small, single chapter documents (e.g. a Quick Reference User's Guide or similar) may require little or no numbering system; even page numbers may be superfluous in some instances.
More complex publications are likely to need multiple numbering systems. As an example, a comprehensive Maintenance or Workshop Manual may comprise more than one volume, and may need a numbering system that controls numbering for:

- The publication itself (product code and/or library reference)
- Modification record (issue numbers, amendment numbers)
- Volumes
- Chapters
- Sections (sub-chapters)
- Pages (preliminary pages often given lower-case Roman numbers; other pages may include chapter number or appendix or amendment number)
- Paragraphs (numbered where individual paragraphs contain procedures etc. to be cross-referenced, or as required by a controlling specification or standard, such as MoD publications)
- Sub-paragraphs (rules to cover sub-sub-paragraphs also, if applicable)
- Figures (possibly different numbering for in-text and 'chapter end' figures)
- Tables
- Appendices and/or annexes.

As a general rule, shorter user documents seem 'friendlier' if numbering – which can appear to be unnecessary 'clutter' – is kept to a minimum. Conversely, larger, more deeply technical publications require fully-structured numbering schemes to reduce the time spent locating wanted detail.

For publications containing numerous chapters, numbering that includes the chapter number simplifies cross-referencing and can avoid confusion. For example, 'See the System Block Diagram Figure 1-3' tells the reader that the referenced figure is the third one in Chapter 1.

Terminology and spelling control

It is important that terminology is consistently applied throughout a range of publications. Other standards may also apply, such as AECMA Simplified English. Inconsistent use of terms or the multiple use of different terms for the same thing will confuse readers and should be avoided by defining common terms in a Glossary included in the Style Guide.

Similarly, spelling standards should be defined – for English language documents the basic choice is between UK English and American English. Whichever is chosen, set your Spell-Checker function to the appropriate one in *all* your word processors, then ensure that the Spell-Checker is *always* used. All these measures can be covered in the clauses of the Style Guide.

Parts of speech and abbreviations

A good Style Guide will cover these aspects, which are impinging on the areas covered in defining 'writing style'. Remember that the introduction to this chapter referred to the essential difference between a Style Guide and writing style. However, in order to minimise diversity of writing styles within the same publication or a family of publications, the Style Guide should include some basic rules, such as:

- Use only the active voice in procedures ('tighten the screw' rather than 'the screw should be tightened')
- Avoid the use of the passive voice in description and operation texts

- Always use a term in full the first time you use it, followed by the abbreviation in parentheses. Subsequently, the abbreviation may be used instead of the full term
- Ensure that abbreviations used comply with the relevant standards. Define those standards in the Style Guide.

Structuring the text

Avoiding long paragraphs and long sentences allows the reader to absorb the information with necessary and natural 'pauses for breath'. Similarly, keeping sentences and paragraphs to cover single ideas and topics also helps the reader to understand the written matter without the need to go back and read the same text again.

The Style Guide should give guidance on these 'writing style' aspects to increase uniformity of publications.

Illustrations and figures

Structure the Style Guide for illustrations and figures to include rules so that:

- Captions on originals are readable at the finished size used in the publication
- All graphics are in portrait format wherever possible (no need to rotate the document 90 °to read the information therein)
- Boundaries are defined for applicable page sizes (including 'foldouts' if appropriate)
- Illustrations and figures are positioned as close to the related text as possible. This may not always be possible, for example, where a controlling specification requires 'foldouts' to be arranged in sequence at the end of chapters
- Illustration/figure titles are positioned in a standard way with respect to the graphic content

- Text is not created as part of the graphic if translation is a likely requirement for the publication. If text is necessary, for example to identify components depicted, use text boxes to contain the text to simplify the translation process
- Line thicknesses and types are defined to have common meanings, for example extra thick lines for system boundaries, dashed lines for external or optional items, etc.
- Exploded views for illustrated Parts Lists are not too cluttered. The usual method is to define the maximum number of items depicted on a given page area
- Items numbered on a figure are numbered sequentially beginning at (say) the top left corner and proceeding in a clockwise direction. This helps enormously in locating specific item numbers
- For different views of the item included in the same Figure, the projection to be used is specified (first or third angle)
- For flow diagrams, the convention of top to bottom and left to right is maintained
- For complex diagrams (for example, circuits), multiple sheets of the figure are linked with appropriate continuation marks or symbols to aid understanding
- Electronic format of figures is readily importable into the text document, is to a defined version of the software, and is in all needed respects compatible with in-house systems in use. See *Specification of Software Tools* later in this section.

Tables

To ensure consistency of style, include rules for tables, covering aspects such as:

- Position on the page (centred between side margins usually looks best)
- Line thickness (perhaps a thicker line for the outside frame and/or between repeated sets of columns)
- Use of shading (perhaps heading line only, or every so many lines to aid navigation)

- Repeat of headings if a table runs to more than one page
- Centering of headings and table data in cells, as appropriate
- Font and size of text
- Position and style of table title.

The following simple table demonstrates some of the above points. Notice that, in this example, the font used is the same as that used for this chapter's body text. The headings have been made bold for added emphasis.

Height (cm)	**Weight (gm)**	**Height (cm)**	**Weight (gm)**	**Height (cm)**	**Weight (gm)**
1.0	50.25	2.5	112.24	4.0	183.64
1.5	68.33	3.0	131.99	4.5	207.03
2.0	94.81	3.5	157.97	5.0	220.93

Height/weight relationship of samples

Headers and footers

Used to help navigation within a publication, define page headers and footers within the Style Guide to ensure consistency of use and uniformity. Information provided is outside the normal page text area. As the names imply, headers are contained in the top page margin, footers in the bottom page margin.

Word processing software allows you to create almost any style you want. Text can be centred, left or right justified (or all three), and can differ for left and right pages, and for different chapters and sections.

Word processing software allows the 'automatic' page numbering to be positioned in the header or footer (more commonly the footer is the preferred location). Page numbers are often positioned centre bottom for single-sided publications, and bottom outside for double-sided documents. Thus the page number for an even-numbered or left-hand page would be to the left side, and for an odd-numbered or right-hand page it would be set to the right side. That convention makes the page numbers easier to view when turning the pages.

Graphics can also be included, for example a company logo. Whatever style you decide on, remember they are intended as an aid to reader navigation; avoid over-complicated schemes that may confuse the reader. The following simplified example shows a typical arrangement:

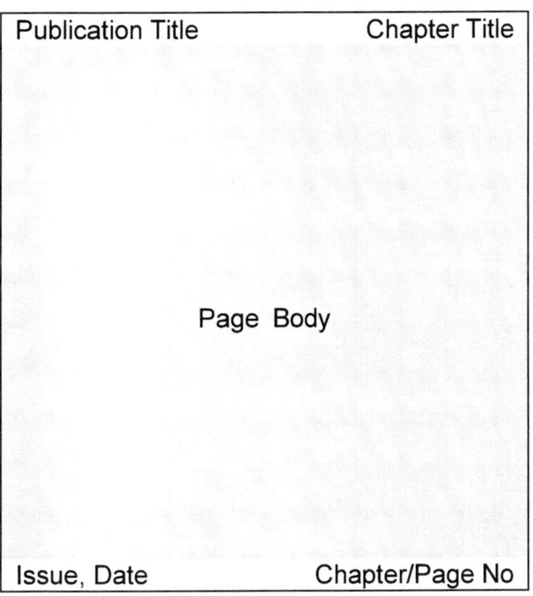

Specification of software tools

In order to ensure complete uniformity and interchangeability, your Style Guide should include reference to specification of software used in the creation and production of your publications.

Documents produced in hard copy may include soft copy as a deliverable to the customer or end-user. It is therefore necessary that the specific software packages (and versions) used to produce the publications are determined before work begins. For this reason, the Style Guide should at least flag up this requirement, which may be ignored if irrelevant for specific applications.

If your publications include examples that are delivered primarily in electronic format, but must include options to print in hard copy, take care that your Style Guide makes allowance for the tools to be used for this specialised but increasingly frequent requirement. It is a fast moving environment which could very easily make your Style Guide out of date if your specification of this aspect is too restrictive.

Specification of fonts

Today's word processing and desktop publishing software packages can offer a bewildering array of fonts, a temptation for the author wanting to make a mark by using an unusual one, or a selection from the huge choice on offer. However, please bear in mind possible problems:

- If a document created using an uncommon font is downloaded and printed on a different system, which does not have that font, the result may be either garbage or a different (default) font which may introduce unexpected characters
- Mixing too many fonts in the same publication can be irritating and confusing to the reader
- Certain fonts may be difficult or tiring to read.

When specifying 'standard' fonts in a Style Guide, remember that most Windows-based word processors will have Times New Roman and Arial fonts. Specification of either will obviate the problem described in the first bullet point above.

The body text of this chapter has been created using Palatino 10 point. Palatino is a font featuring serifs on the font characters, and is considered to be easier to read than a similar-sized sans serif font such as Arial.

This paragraph has been created using Arial 10 point. The difference between the two fonts is quite marked. As mentioned earlier, Arial is widely used.

Except for publicity brochures and similar documents, where font choice may be influenced by 'artistic' or corporate identity reasons, the general rule is to severely limit the number of different fonts used in the same publication. However, you may, for added emphasis, want to use a different font for, say, WARNINGS and CAUTIONS.

Earlier in this chapter, you will have seen an example of a hierarchical headings structure. This was created using different subsets of the same font, and is easy to understand, to recognise and to follow. If such a structure were to be created instead using a selection of different fonts, it would be much more difficult to remember and could make navigation within the publication less easy.

Font size is also worthy of consideration when setting the rules in your Style Guide. The size used here is generally considered to be readable without effort for most people. A larger point size would be even easier, but would increase the page count of the document and have the likely effect of causing text on a single topic to run to more than a single page. Conversely, too small a point size reduces readability. In the extreme, a very small font size will discourage users from reading the document at all.

A final word on choice of fonts. Your company's standards (or your client's standards) may make the choice for you. Check the likely present and future requirements before finalising this part of your Style Guide.

Section 4 – Document structures

General

This section touches briefly on the structure of publications, in terms of chapterisation, use of annexes and appendices, etc.

Chapterisation

Breaking a large document into separate 'blocks' called chapters is familiar to everyone; just about every novel you care to take from a bookshelf consists of a number of chapters to make up the whole story. Why do authors do this? One answer is to break the whole into smaller 'bite-sized' chunks, making it less daunting to the casual reader, much as was mentioned earlier in this chapter when discussing paragraph breaks and line spacing.

Another reason for the technique is that it is a way of conveniently separating areas of the story. An example is the type of novel that tells parallel tales about different characters. The chapter break then forms a convenient point to end an episode about one character or scenario and begin the next episode about another character or scenario.

In technical publications, both of these purposes have some validity. The reader or end-user is less daunted by a series of clearly separate chapters than a hefty book which appears to be unbroken from first page to last, and can navigate more easily when chapterisation shortcuts the locating of wanted information.

Probably the major reason for splitting up a technical publication in this way is for ease of navigation. End-users, who may be product users, maintainers, etc., will locate wanted information much faster if there are clear navigation points, or 'landmarks'.

Chapter breaks are obvious landmarks, particularly in the type of publication, which utilises tabbed divider cards between chapters. Divider card tabs may be plain, numbered (or lettered), or even printed with the chapter title. Sometimes multi-coloured tabs are used as a further speed of access feature. The user of the publication can simply turn to (say) the red tab for the fault-finding chapter.

Other methods of highlighting chapter breaks include colour-coding the pages themselves, or applying a coloured strip down the outer edge of each page. In black and white documents, a similar scheme can be employed which uses 'finger tabs' giving the chapter number or title, printed on the outer edge of each page. The vertical position of the finger tab is different for each chapter, and staggered through the document just like a set of physical divider card tabs. Figure 6-2 shows all the tabs for a publication; you would of course see only one of them on any individual page.

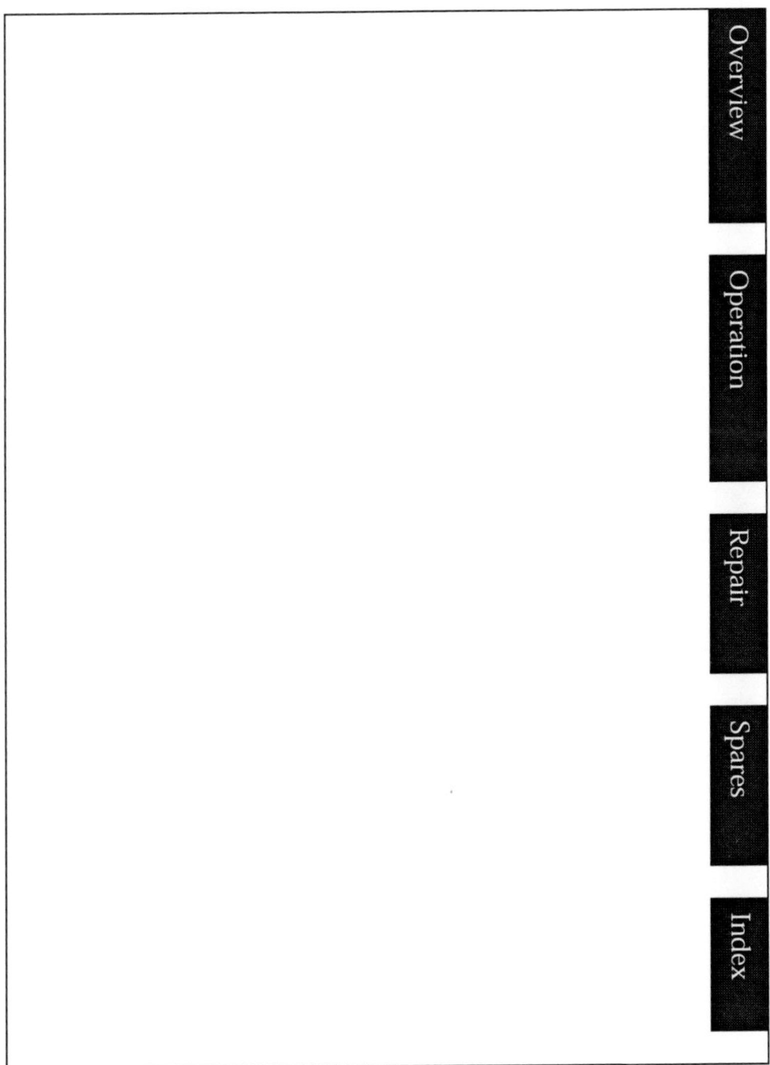

Figure 6-2 Example of finger tabs for document navigation

Appendices and annexes

We have all seen publications with appendices or annexes after the main body of the document. Your Style Guide should not only define the numbering scheme to be used for them, but when they should/should not be used and what they should contain.

A suggested definition of an appendix is: 'A repository for supplementary information that is needed by the reader, perhaps for reference purposes, but is not appropriate to include in the main body of the document'.

Typically, appendices may contain specifications, error messages, sample programs, practice exercises, training records, and similar.

Some Dos and Don'ts for your Style Guide:

- Do begin each new appendix on a right-hand page
- Do give each appendix a title and a number or letter (and make sure the Contents List includes them)
- Do give the pages distinctive page numbers. As an example, the first three pages in Appendix A could be numbered: A-1, A-2, A-3
- Don't be tempted to dump information into an appendix that is of no real use or interest to the reader, for example deeply technical information that a design engineer thinks should be included
- Don't include the index in an appendix.

Sequence of information

If your Style Guide includes specification of one or more 'standard' publication types, it may be possible to stipulate the order or sequence of the information contained within a given manual type. Standardisation of chapter order in similar manuals produced will help end users.

When generating a specification, try to make the sequence logical for the intended end user. In User's Guides for example, beginning with a 'How to use this Guide' before the main body of the document will help readers and give them added confidence. Although individual requirements will be project or product dependent, a typical breakdown for a User's Guide might be:

- Preliminary pages including contents, glossary, associated publications, section on safety, etc.
- How to use this guide
- Getting started
- System/product description
- Operation (basic)
- Operation (advanced)
- Troubleshooting guide
- User maintenance
- Index
- Appendix: On-screen error messages.

For a maintenance manual, aimed at a different end user, the corresponding information sequence might be:

- Preliminary pages including contents, glossary, associated publications, section on safety, etc.
- Introduction/overview
- System description
- Installation information
- Fault-finding guide
- Recommended test equipment
- Adjustments, testing and repair
- System drawings
- Parts/spares lists
- Appendix: Preventative maintenance schedules.

Similar lists can be compiled for each type of publication to be covered by the Style Guide, customised to comply with the requirements and products of the organisation involved.

Flexibility of rules

Wherever possible, allow flexibility in the Style Guide rules. You may for example find that an individual manual requires the inclusion of very large, complex drawings that will not fit in the selected page format (even as foldouts) and would; if reduced, contain text information far too small to read. Obviously a different solution would be called for.

If the drawings could not be redrawn and split into several sheets, another possibility would be the use of associated binders with the large drawings, multi-folded to a final (say) A4 format.

Alternatively, you may consider having an associated volume at (say) A3 format, to avoid handling problems for the maintainer, who may otherwise be struggling with the many flapping folds of drawings in hostile environments.

Section 5 – Illustrations and other graphics

General

It is undoubtedly true that good illustrations and graphics contribute enormously to the understanding of the content of a publication. Very few manuals would succeed without 'pictures'. However, there has to be consistency of style, positioning, numbering, and so on, if the finished publication is to look good and to give maximum value to the user.

Where to use them?

In many publications, the first 'picture' encountered by the reader will be a Frontispiece illustration. Dependent on the subject matter contained, this could be a system block diagram, a photograph of the product, an illustration of the product, and so on. It is just as important with graphics as it is with the text to bear the end user in mind when determining the style of 'picture' to include.

As a general rule within a publication, match the style of Figures with the associated text. For example, a detailed system block diagram would be out of place in an introduction or overview chapter.

However, your Style Guide will be more likely to be specifying other factors, such as:

- Position – usually as close as possible to associated text reference, centred between side margins
- Size – including whether whole page only and/or in text
- Numbering scheme – none, chapter-based or publication-based
- Title styles and positions
- For electronically produced figures, the software and version.

For User's Guides describing application software, it is increasingly the case that figures showing screen captures, and used to illustrate procedural information, are not titled or figure-numbered. You may wish to title or number them in the traditional way in your Style Guide.

You may also wish to specify that electronic drawing files are linked to the publication, but not embedded in it. This technique gives smaller text files that can be handled more easily.

Deciding on figure types

Just as with the Frontispiece discussed earlier, the subject of the publication will influence the choice of the most suitable type of figure to be specified in any instance.

For example, quoting an applications software User's Guide again, it is less likely that this will use many complex line drawings. In this type of publication, you will want to see more screen captures, graphs and charts, simple algorithms and photographs.

A typical Maintenance or Repair Manual may in contrast require many schematics, exploded view illustrations, and in-text adjustment or setting figures. The maintainer will require the fine detail of illustration to effect maintenance and repair of the system or product.

It is common practice in such large and complex manuals to have the majority of the drawings grouped at the end of each chapter, in many cases as foldouts. This system overcomes frequent interruption to text flow that would result if the figures were placed close to the text reference. It also gives a finished manual that is easier to print and collate.

A repair manual often includes an illustrated parts list. Here again, exploded views will be used, annotated with item numbers cross-referenced to adjacent parts lists. Some illustrations may need to be repeated to always face the parts list page that includes the item numbers on the drawing.

All these factors can be considered when compiling the rules for figures in your Style Guide.

Some figures Dos and Don'ts:

- Do keep styles and sizes consistent
- Do position in-text figures close to the associated text
- Don't use a complex drawing reduced to the point where captions are unreadable.

Section 6 – Style guide aspects for on-line documentation

On-line/paper differences

If your Style Guide is to include the specification of on-line documents, remember that:

- On-line documents are displayed in an area that is smaller than a paper page, and has different proportions
- Whereas most paper documents are 'portrait' format, monitor and TV screens have a 'landscape' orientation
- A standard monitor may typically display as few as 24 lines of 80 characters – about 320 words for a full screen of text. This is far less than a paper page, which may contain up to 60 lines of text
- One of the disadvantages of the on-line presentation of documents is that users tend to follow instructions blindly if they form part of a long procedure, but only a small section of it is visible on the screen. This means that navigation and landmarks are more important
- Tests have shown that most people prefer a document which has a width approximately 1.6 times its height. This may seem to conflict with the fact that most paper pages are taller than they are wide. However, we do usually view a two-page spread when looking at a paper document. Considering A4 pages as an example, a two-page spread has a landscape aspect ratio of 1.41 to 1
- Negative contrast (dark characters on a light background) has been found to reduce errors, to speed up work, to increase comprehension, and to improve reader comfort and legibility

- Resolution of even high-resolution screens is low compared to typically good quality laser-printed paper pages. Essentially, you cannot display on screen, graphics with features smaller than a single pixel
- Colour rendition on a screen may appear different, and may comprise a limited range of different colours on certain monitors.

Planning your screen layouts

Taking all these factors into account, you should design your screen styles to define what can and cannot be displayed, where on the screen area features will be positioned, and how they will be implemented. As a starting point, try to create a plan in a methodical way, considering points such as:

- A list of what needs to be displayed; items may include:
  - Contents lists
  - Document title and other headings
  - Text and graphics
  - Navigation buttons or links, etc.
- Organisation of groups of similar items and of relative priorities of displayed information. Consider use of 'attention grabbers' such as blinking text or different colours
- Layout of the display so that users can find wanted text and other information effectively and easily.

Having made a plan after considering these and any other factors, be prepared to revise your plan until you are satisfied it fully reflects your anticipated requirements.

Split screen systems

Some successful on-line documentation screen systems use a split screen, where the major part of the screen contains the document 'window', and the left side is a separate 'window' containing a fixed or scrolling contents list in the form of a permanently visible navigation aid.

Figure 6-3 shows a representation of just such a system, which helps overcome one of the major problems of on-screen documents – that of getting 'lost' in the middle of a document.

In the example depicted, the part of the publication in view is indicated automatically in the left segment of the screen by highlighting in the 'contents list'.

You can specify the set-up so that by clicking onto a heading line in the left-hand part of the screen, the main screen view will take the user to that point in the document.

Some on-line Dos and Don'ts

- Do design your layouts carefully; users should find the screens easy to use and restful to read
- Don't go overboard with colours; just as proliferation of fonts in a paper document is confusing, too many different colours are also to be avoided on screen
- Do use ALL CAPITAL LETTERS sparingly; all upper-case words are slower to read by around a third, compared with the same words in lower-case
- Do keep line spacing sufficient to make reading strain-free, but avoid too much space, which limits the amount of text that can be viewed without scrolling
- Do try to keep line lengths short. Longer lines require more eye movement and can be tiring
- Don't justify text. Just as for paper documents, unjustified or 'ragged' text is easier to read
- Don't use "As shown above" or similar; this can irritate users who can't instantly see the item referred to
- Do specify good Help and Find systems; easy navigation is vital in on-line documents
- Do include graphics wherever possible, but bear in mind the limitations of resolution on-screen.

Do remember that complex graphics, especially in colour, can increase file sizes dramatically, slowing down movement around the document. A 'trick' is to call up a graphic from a stored location, allowing the same graphic to be used several times without increasing file size.

Chapter 1 Overview
General
System Features
Hardware Options
Main Unit
PCBs
Software Options
Standard
Custom
Chapter 2 Description
Introduction
Main Unit
Processor Unit
PCBs
Power Supplies
Plug-ins
Expansion Slots
Chapter 3 Operation
Getting Started
Basic Operation
Advanced Operation
Supervisor Features
Chapter 4 Troubleshooting
General
Alarm Indications
Status LEDs
Error Messages
Power Supplies
Chapter 5 Maintenance
Preventative Maintenance
Corrective Maintenance

Introduction

The equipment comprises a Main Unit, external interface accessories, and the application software, which defines the functional behaviour of the equipment.

Main Unit

The unit is configured for 19-inch rack mounting, and comprises a Main Unit with Plug-in Printed Circuit Boards (PCBs) in a full width card frame, in which the card front panels are forward-facing and are visible through a transparent dust cover forming the unit front panel.

Power Supplies are provided from a Switched Mode Power Supply (SMPS) contained within the Main Unit and mounted at the rear of the card frame.

Dual, thermostatically-controlled fans dissipate excess heat generated while the unit is operating.

Processor Unit

The Processor Unit housed within the card frame in the Main Unit contains the PCBs, plugged into the

Figure 6-3 Typical split screen layout arrangement

Section 7 – Configuration control

Before you issue your Style Guide, plan in future amendments and revisions, and how you will ensure that:

- You will know who has copies of the original issue
- Every copy holder will be certain to receive any updates and that you will know they are received and incorporated.

The Style Guide should help ensure that all your publications are consistent in specified style aspects, but this goal will not be achieved unless all involved personnel work to it, and always have the latest issue.

The necessary measures and rules to ensure this happens can be built into the Style Guide itself.

Single-point control of the issue and distribution of the Style Guide will avoid the existence of uncontrolled, obsolete copies being used in error, creating out of step versions in circulation.

Signed acknowledgement of receipt of updates will help control issue status of copies. Spot inspections to follow up will detect misuse of the configuration control system, such as failing to incorporate received revision packs.

7
Illustration (information-graphics) for technical publications

Peter Lightfoot, FISTC *has spent a career within technical illustration and information graphics, which spans more than four decades. It also spans the period that included the high point in traditional technical line and tonal airbrush illustration through the transition to digital illustration and desktop publishing. This career followed a mechanical engineering apprenticeship, and subsequently led to work in the aircraft, shipbuilding, petro-chemical, oil & gas, machine tool and automotive industries carrying out tasks that cover all aspects of information graphics. He is currently a founder and director of Media 4 Graphix Ltd who specialise in the production of conventional and on-line technical and commercial promotional material.*

Info-graphics is a new name for a profession that still requires the application of traditional drawing and illustrative skills

"Illustration can be defined as the creation of a specific visual image, for publication, that universally offers immediate communication to the reader (having the ability to inform, identify, instruct and help visually relate to details contained within the subject) in support of a given text."

Introduction

This chapter places the use of visual renderings - both figurative and abstract - into the context of technical and scientific informative and instructional communications. Their use extends from the illustrations within conventional printed publications to digital renderings for on-line documentation. Both require the same skills and ability to interpret the author's requirements. No matter the media, such instructive documentation will always require some form of graphical interpretations to assist the reader to fully understand the written word.

The practice of information graphics, or technical illustration, to depict technical subjects has been refined over many centuries, culminating in the present styles and techniques, which deliver the clarity and consistency required for mass communication.

Illustration is the only means of communication that is truly international, through its ability to communicate across the barrier of language. The old cliché still pertains, "that a good picture is worth a thousand words".

The assimilation of many detailed descriptions, especially for the most complicated of technologies, is more readily achieved if supported by appropriate graphic rendering. The experienced illustrator has an armoury of visual interpretative techniques from which to choose the style of illustration that best supports the author's copy and enhances the reader's ability to understand.

In more recent times, the environment in which illustrators operate has changed dramatically. The production of the graphic interpretation has moved from one of using traditional artistic hand-skills, to one using computer-aided technology; and the illustration reproduction from photomechanical means, to one that is almost totally digitally produced.

However, the role of the technical illustrators and their objectives remain fundamentally unchanged. Computer technology has only changed the tools and working methods and because of its automated techniques it has affected graphic production styles for good, but the need for illustrations and the information they impart still remains.

Line illustrations - originally drawn manually on a white board by the traditional method of pen and ink - have been superseded by the use of vector graphic software that digitally reproduces the same forms of illustration – information graphics.

Virtually all the publishing industry now operates using digital reproduction systems in whole or in part. This has led to the acceptance of digital artwork as a standard rather than the exception. All forms of artwork - line, tone, photo; mono or colour images - are now converted to a digital format, before being printed or presented on-line. The digital scanner has become an essential tool of the illustrator to prepare graphic outputs that comply with publishing and on-line formats.

Having said that, an illustrator still requires to be trained to obtain a complete grasp of traditional methods and techniques for the production of acceptable representations - a full grounding in drawing techniques, composition, perspective and colour theory. They need the practical ability to be able to render figuratively what can be seen physically as well as the skills to interpret complex assemblies from detailed component and assembly drawings obtained from technical sources such as draftsman's elevations or CAD outputs. It is still essential that these skills be mastered in the traditional way, before transferring them to the digital means of production.

Today's illustrator, beside being computer literate and having interpretative graphic competence, also requires an enquiring mind and good personal communication skills. They need these in order to work with the author and support the research and investigation which may be necessary to deliver an optimum graphic statement.

In transferring the traditional manual graphic skills to digital technology, it is interesting to note that much of the traditional terminology such as brush, pen, eraser, airbrush, cut, paste, trace, render, and registration has been maintained to describe the functions of the programs. As a traditionalist, nothing can replace the smell of the adhesive and the danger of the scalpel.

Many graphic programs now exist to reproduce line and tone renderings: three-dimensional views which can be cutaway or exploded, isometric and axonometric rendering, or basic two dimensional elevations, abstract schematic or graphical diagrams in 2-d and 3-d formats. No matter how comprehensive and user friendly these specialist computer programs may be, it still requires a skilled graphic interpreter, the illustrator, to select and produce the optimum viewpoint for maximum understanding.

Illustrative rendering - application, styles and techniques

Illustrations fall into two major categories - figurative and abstract.

Figurative illustrations are those used to realistically represent a subject, in either two or three dimensions, in a configuration which emphasises and aids the interpretation of the written description. This form of illustration also includes photographic images that can be cropped, retouched or manipulated by the illustrator.

Abstract illustrations are schematic or diagrammatic renderings, which can also be produced in either two or three dimensions, and are used to represent theoretical subjects or processes. They are used to depict measurements (quantity, distance, time, speed, temperature etc.) and control or functionality (direction flow, power sources, etc.). The most usual forms of abstract illustration are graphs or charts such as pie, column or block charts, schematic diagrams for circuits, or blocks that represent active components within a system or process assembly.

The techniques used for traditionally or digitally rendering figurative or abstract illustrations for publication fall into two categories - line or tone.

Line drawings

The term 'line drawing' refers to any drawing or illustration that consists simply of black lines on a white background, or the reverse. In other words, every mark and the paper or material on which it is made has a black to white contrast, without variation in tone as in a photograph, airbrush, pencil or wash drawing. In a line illustration, the form or shape of the subject, surface textures or tonal-variations may be indicated as thin lines or dots. Within these limits there is scope for a wide variation of choice.

One of the main advantages of the line drawing is that it can be easily reproduced, by direct copying methods, and printed. There are many categories of line drawings as explained below, and you should choose the most suitable for all the purposes required.

Continuous tone illustration and photography

In tonal techniques, the image is produced by the application of an infinitely variable range of tone from dark to light on the board or media. The range goes from black though tones of grey to white for monochrome illustrations or equivalent tonal values from basic colours, for full chrome illustrations.

Typical of continuous tone illustrations are photographs, or drawings rendered by the use of airbrush, wash, and pencil or charcoal drawings.

In this case the illustration is treated as a photograph and screened for reproduction. It is possible, however, to give a line drawing tonal or shaded effects by the use of mechanical tints, but in this case the illustration still maintains its line characteristics, and screening for reproduction is not necessary.

A high-quality half-tone illustration can be a desirable asset for publicity or promotional material. An expert airbrush artist can create a photographic impression of an object before it has been manufactured, but the comparative cost of its production is high in relation to line mono renderings.

To publish this type of illustration requires a photographic or digital scanning process to translate the image into a series of tiny dots varying in density according to the different tonal values of the image. The processed result is referred to by the printer as a half-tone.

Colour

Most illustrations and diagrams can be made clearer by the use of colour, especially those showing flow paths of gases and fluids. However, before the advent of digital and on-line publishing the use of colour in technical documentation was generally not economically viable. Now new production and printing technology has made the use of colour a cost effective option for the illustrator to employ.

It is recommended that its application in information graphics is best employed by those trained in its controlled and aesthetic balance.

Figurative illustration techniques and finishes

Two dimensional illustration - orthographic projection

An orthographic drawing can often be the most suitable way of illustrating the subject, because in most cases it can be taken directly from the manufacturer's production drawing. Orthographic projections can be a very cost effective form of illustration. However, such illustrations are still required to be prepared by an illustrator to ensure they meet the requirements of the documentation.

The main limitation of orthographic drawing is that it is only two dimensional (height and width) giving a flat view on one plane of an object and therefore only details on that plane can be shown.

This can be overcome by including a view from another direction, either complete or in part (Scrap view), depending on what hidden details need to be shown. Alternatively, if the illustration has to show details on every side of an object a conventional three view general arrangement can be used. Such a drawing can also be used to convey overall dimensions for installation instructions or to show clearances for servicing manuals.

Sectional views

Probably the most effective orthographic illustration is the sectional view, which shows the inner workings of the subject. It can be used to show the principles of operation and flow paths as well as the locations of the working parts. It can be used to support descriptive text, using direct annotations, and again as a spare parts illustration using item numbers. This can be achieved either by taking a secondary master from the original before annotating, or by using the original with two clear acetate overlays for the numbers and annotations. The latter means there is only one master to modify if there are any changes in the equipment. The disadvantage of a sectional view for a spare parts illustration is that the items are not isolated as in an exploded view. But on the other hand it is generally easier to see the functional relationship between items on an assembly.

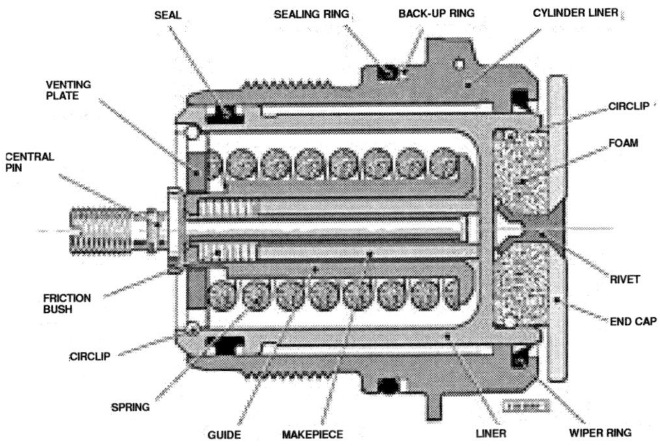

The main advantage of an orthographic section is that it can cost only a fraction of the figure for an equivalent perspective cutaway or exploded view, and generally the research involved is much less than for a perspective illustration. The reference material must be clear and easy to follow. The magical powers that illustrators are supposed to possess that enable them to conjure up details out of the blue, under the heading of artistic licence, are very limited in a sectional view because of its functional nature, and a good copy of the original must be available. The illustrator should prepare an orthographic view in a style consistent with that used throughout the publication and within the limits of interpretation by the end user.

3-D illustration methods - metric projections and perspective techniques

The three dimensional illustration is the main part of the illustrator's craft. To plan and execute an illustration through all its stages, from initial construction to final inking is a very satisfying task, and the result can be fulfilling if not always financially rewarding. However, the cost and time involved must be justified.

Although in one or two cases the metric projection has the advantage, the well-planned perspective drawing has the ability to present a complete, optically correct, visual description of the subject. In one illustration it is possible to show the overall shape or profile of an object, as well as inside details and inner workings.

Metric projections are set up from orthographic projections and can be drawn to scale.

Isometric projection

The most suitable and reasonably realistic perspective illustration is the Isometric projection. Base lines of the object are placed at 30 °to the horizontal, and length, breadth and height are each drawn to actual scale. All circles appear as regular ellipses (35 °-18 °) to whatever scale, and all straight lines can be drawn with a 30 °set square.

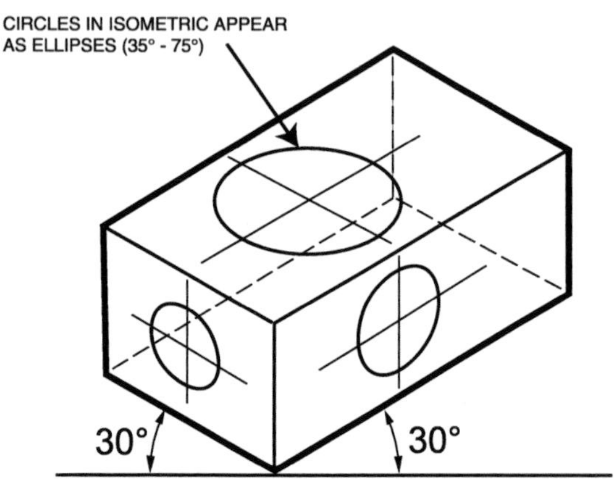

ISOMETRIC PROJECTION

Axonometric

An Axonometric projection is a similar form of projection that has the added advantage of containing a true plan view of the object and is therefore more easily set up from existing drawings. It is particularly suitable for the diagrammatic view of building interiors. For convenience this projection is normally prepared using the base lines length and breadth at 45 °to the horizontal, but it can equally be prepared at 30 °or 60 °.

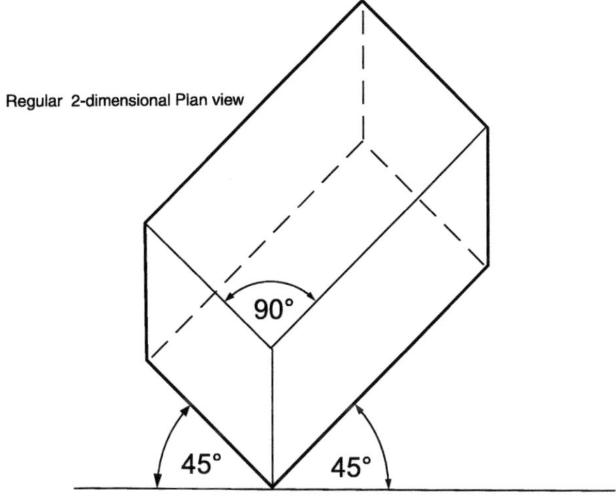

AXONOMETRIC PROJECTION
(USING 45° - 45°)

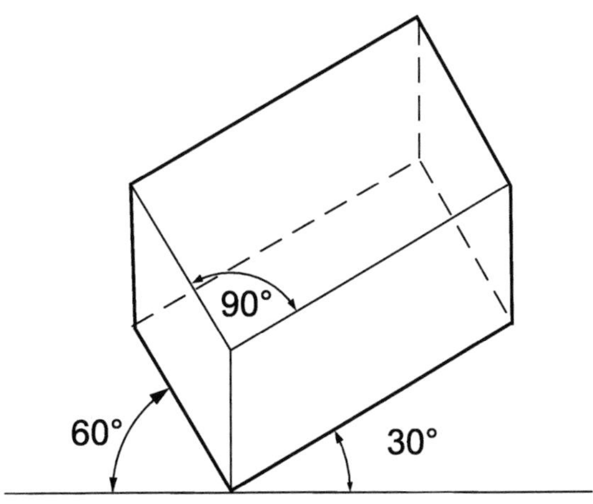

AXONOMETRIC PROJECTION
(USING 60° - 30°)

Measured perspective

Measured perspective is initially the most complex of the projections to use but gives the most optically realistic rendering of a given object. It relies on the projection of all parallel lines to calculated or estimated vanish points located on the picture plane.

Types of view

Views can be rendered in any of the 3-D techniques described, in the following four categories:

Overviews and general arrangement illustrations

The external perspective is ideal for external views and locating external details, such as controls, fixings, gauges and access panels. In cases where items are located on the unseen side of the object an inset detail can be included on the main illustration.

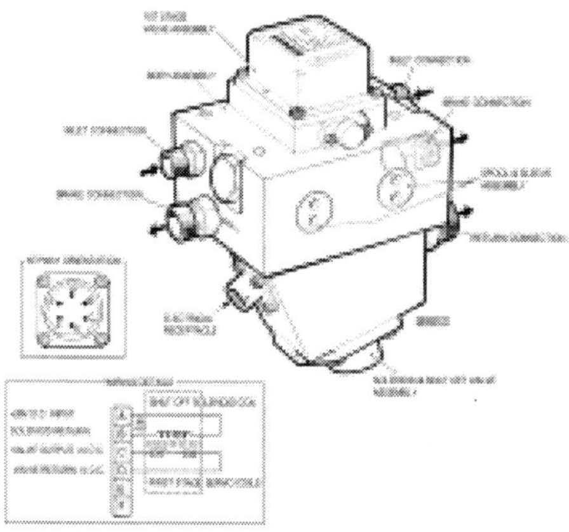

Cutaway illustrations

The cutaway perspective is fundamentally a three dimensional sectional view. It is particularly suitable for subjects that are basically cylindrical or have a central axis, such as pumps, valves and filters, and the usual practice is to cut away a quarter segment to expose the inner details while still maintaining the external shape. Of course, the technique can be applied to any subject, and the cutaway portion can be any shape or size, providing it is clearly presented and the object is still recognisable. As with the orthographic section, this type of illustration is particularly suitable for showing internal flow and principles of operation.

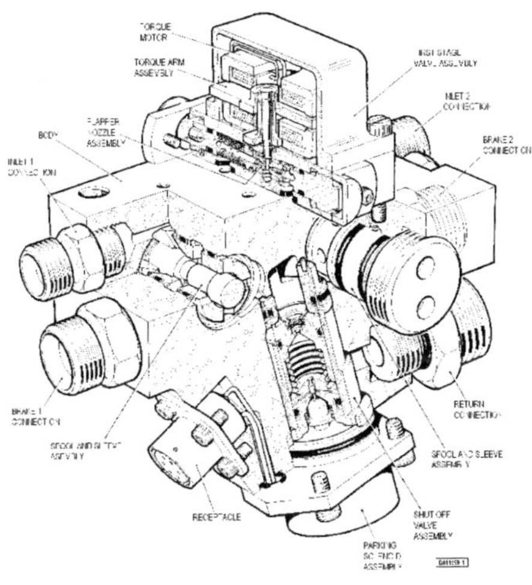

Exploded illustrations

As the term implies, the exploded view shows an assembly with its components blown out from the main body, but still remaining in relative order along their respective centrelines. This type of illustration can be used for a parts list where items need to be isolated from each other while still maintaining their sub-assembly groups. It is also ideal for showing the order of removal and assembly of components.

The exploded view can, however, present a problem of space if the subject is complex and has many items or if the main body fills the drawing area even before the items are pulled out. This can sometimes be overcome by staggering the centreline in a 'Z' formation, or by slightly overlapping each item, but not to the extent where they become unrecognisable. Care should also be taken not to make the drawing too fragmented so that it is impossible to see how the various sub-assemblies fit together.

Illustration (information-graphics) for technical publications

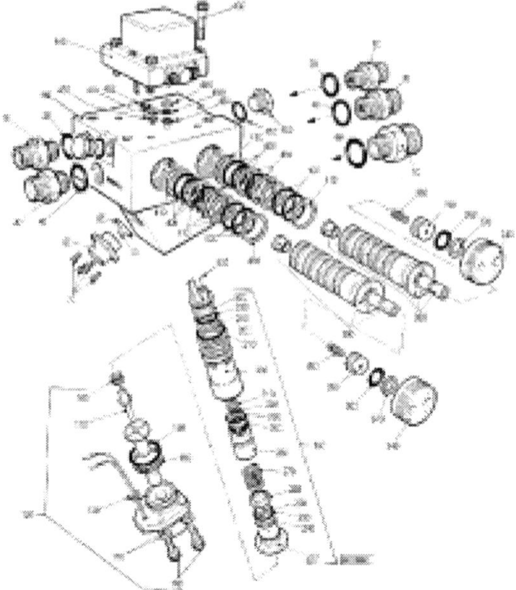

Typical three-dimensional exploded view for a parts listing

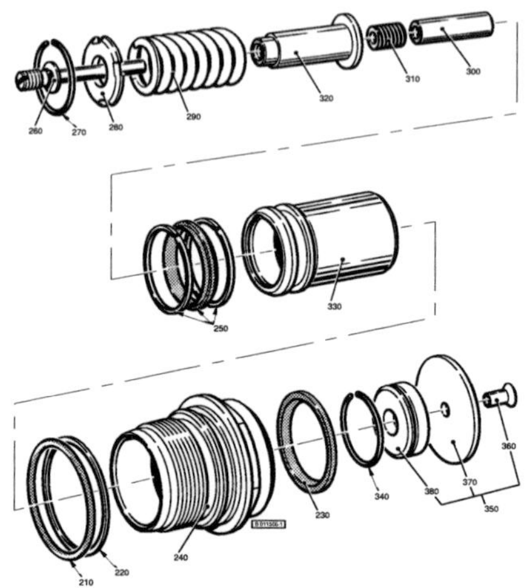

Exploded parts on a common axis

Product visualisation

There may be times when an illustrator may be called upon to prepare an illustration of equipment that is still a figment in the designer's mind. In such cases the ingenuity and technical knowledge of the illustrator will be stretched to create a typical working environment for the subject.

There are now CAD packages that can automatically assemble component details and then rotate the assembled information but the images these packages create are limited in their ability to be used for publishing. They are also very expensive to produce both in the overhead cost of the equipment and CAD operators.

Photographs, photography and photo-manipulation

Photographs have always been part of the illustrator's armoury. Their use has required skill and reproduction methods which have made them not economically viable for most technical publications.

Unless set up specifically as a studio picture - most location and general photographs contain more information than is necessary to support a given description. This requires the finished result to be retouched to enhance the subject.

New digital cameras - with highly sophisticated automated focusing and exposure control now make it possible to take pictures under severe lighting restraints. This coupled to the ability to download images directly to the computer screen, have made their use more economic, although the process of manipulation still requires a skilled graphics operator to adapt and prepare the image for publication or on-line use.

Retouching techniques

Traditional retouching was a technique used by illustrators to adapt full-tone photographs for publication. This involved removing unwanted blemishes, shadows, reflections, ghosting or removing backgrounds. It required a specialist retouching artist - a highly skilled technician using masks and an airbrush. Today these tasks are carried out using photo-manipulation computer programs which are capable of all the aforementioned tasks with many additional features which, in the hands of a skilled illustrator, make colour matching and blending a far more accurate and economic process.

Abstract illustration techniques and finishes

Abstract illustrations are schematic or diagrammatic renderings, which can be produced in either two or three dimensions, and are used to represent theoretical subjects or processes. They are used to depict measurements (quantity, distance, time, speed, temperature etc.) and control or functionality (direction flow, power sources, etc.).

System diagrams

If an engineering manual covers one or more electrical, pneumatic, hydraulic or similar systems, then usually the most important pages in the book will be the system diagrams. These are the pages that will be consulted first in the event of a problem. Well laid out, comprehensively annotated system diagrams tell an engineer a lot about how the system works, how it should be operated and what faults are likely to occur. It is therefore essential that a great deal of thought and care be directed towards achieving the highest possible standard of system diagrams.

The prime responsibility for determining the form and content of system diagrams must rest with the author. Nevertheless, a good illustrator working in close co-operation with an author can be more than just an embellisher of the author's rough layout. If the illustrator follows the logic of the layout, he may be able to spot errors or omissions, or be able to suggest improvements to the layout. There are four main forms of system diagrams:

- pictorial functional diagrams

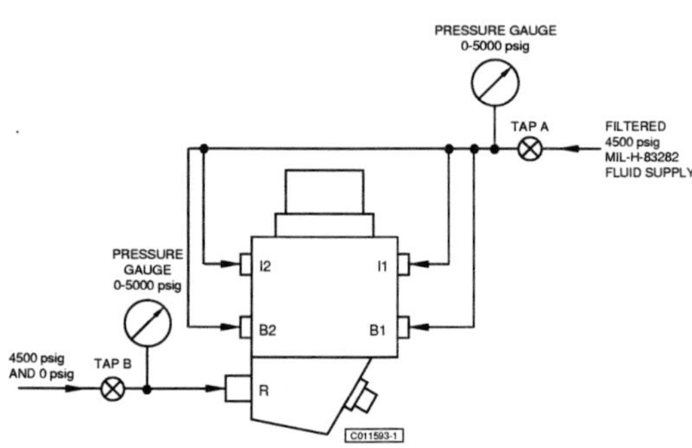

- block functional diagrams

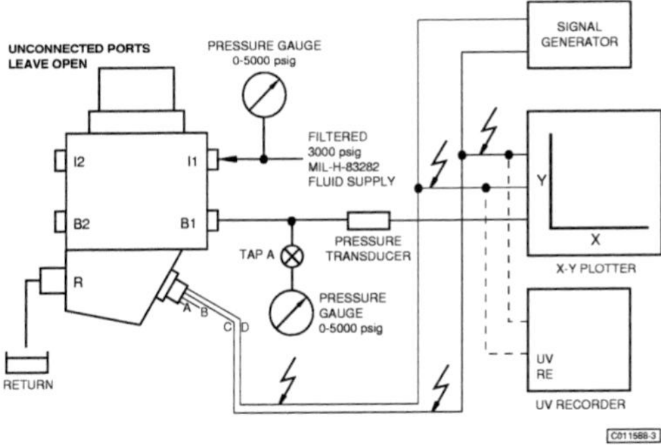

- theoretical circuit diagrams

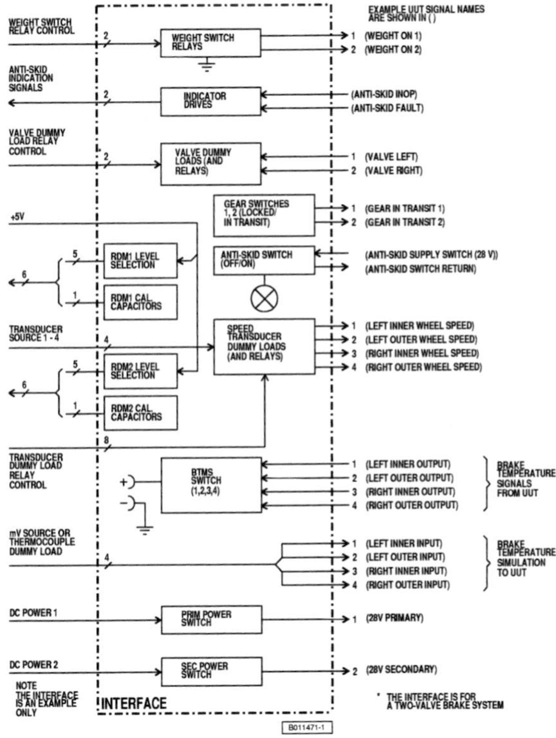

- physical system diagrams showing the physical locations of components and their interconnecting wiring and piping

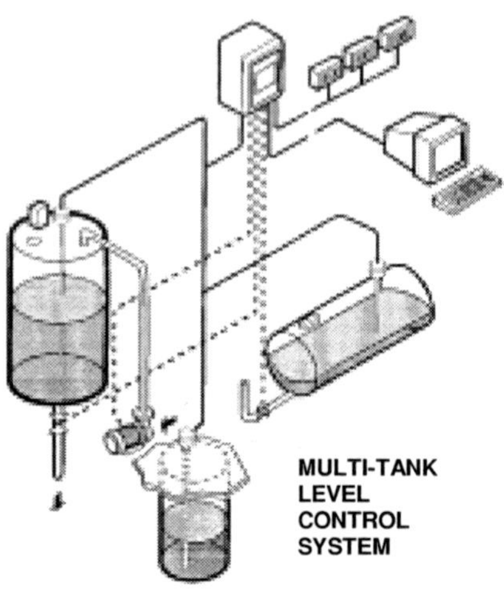

The four types of diagram perform different functions, and unless the system is very simple it is best not to try to combine more than one function on one diagram (for example, by endeavouring to show the physical layout on a theoretical circuit).

Charts and graphs

Graphical representation of statistical information is constantly being used, and examples such as political trends, rise and fall of inflation and how government expenditure is apportioned, can be seen almost daily in one form or another. A well thought out graph or chart can instantly convey the relevance of facts or figures that may otherwise be tedious and difficult to follow.
This type of information can be divided into three categories:

- trends and movements
- percentages and quantities

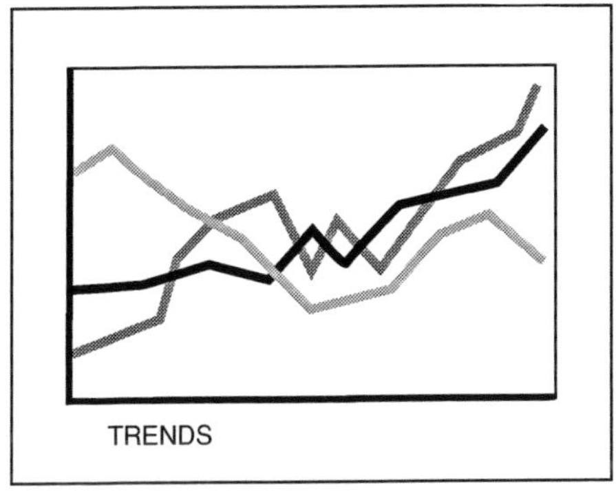
TRENDS

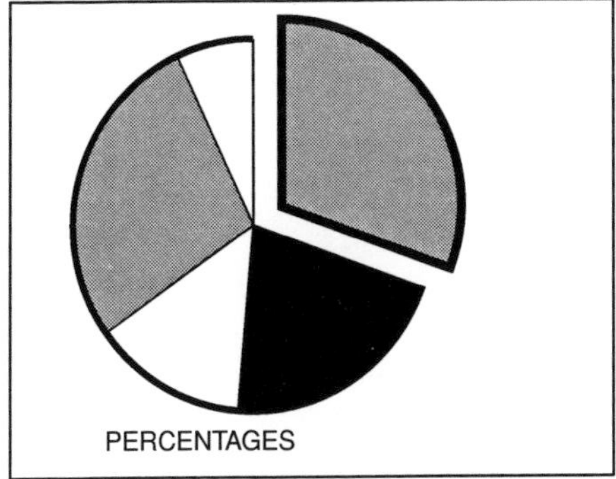
PERCENTAGES

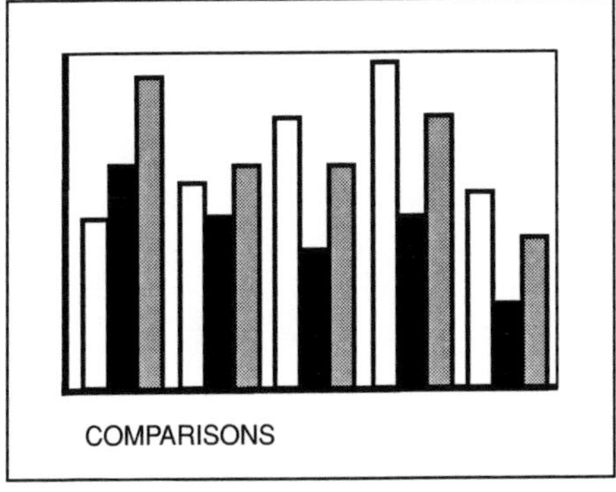
COMPARISONS

Graphs

Simplicity and clarity should be the aim when drawing graphs and charts, and there are many graphical aids available today to help put over the message with the greatest possible effect.

Trends or movements such as sales figures or temperatures, can usually be best shown as a line graph.

Graphs for publication should be as simple as possible, depending on how precise the information has to be. If a grid is used the lines should be kept to a minimum and preferably be a different colour, but this of course depends on the budget for the publication.

Pie and bar charts

Pie or bar charts can best represent percentages and quantities. A pie chart is a circle, or pie, divided into segments to illustrate portions of a whole. For instance, if the whole pie represents 100%, then a quarter segment would represent 25%. The bar chart shows quantities or divisions of amounts by bars of varying length.

Column charts

Comparative quantities or groups are best shown by column charts, in which amounts and quantities can be compared to each other by varying heights of columns. Similar comparison can be made for masses that can also be represented by symbols. For example, bags of money, barrels of oil, groups of people, etc.

Economics

Making a choice

At the planning stage in the production of any sort of documentation, consideration must be made by the communications team as to the form information graphics are to take. In the majority of cases the cost of the illustrations is a significant factor in the overall cost of a publication. In such cases it is advisable that the graphic department is represented from the first day of planning.

Many of the initial questions to be asked about the illustrations will be the same as those for the text, and a close liaison between author and illustrator can result in valuable saving of cost and time.

Where documentation is produced to an industry specific specification, such as ATA100 or DefStan, for its layout, then many of the stylistic questions and pictorial consistency will also be defined.

However, where no fixed format has been decided, a proposal should be created to ascertain a budget for each facet of the publication. One of the most important factors in planning is consideration of the graphic costs.

It would be an ideal situation if the cost of each illustration in a book could be considered on its own merits, and in some cases this may be so, but as most technical documentation contains several sections, each requiring similar renderings, consideration needs to be given to the multiple use any of the illustrations. It is possible to plan the use of such illustrations from the outset in order to reduce cost and improve efficiency.

Photographs or drawings?

There are many pros and cons to be considered when weighing the merits of photographs against drawings. If a good photograph of the subject exists and covers the details required, then it may be considered. But, take into account the cost of preparing it as a half-tone illustration and whether the equipment is likely to be changed in any way. Often modifications are made to a document just before publication. It may be possible to convert a photograph to a line drawing quite simply by tracing off an enlarged print. If photographs are to be taken, then the following thoughts should be considered.

When will the equipment be at the build stage required? Every item needed to be seen should be fitted when the photograph is taken, and the equipment cleared of any extraneous items that may confuse the reader, such as test gauges, wires, fitter's tools and tea mugs. Doing so will avoid the need for expensive retouching or picture manipulation.

A major consideration when working on large projects is how long the equipment be will be available. Remember it takes time to set up a photographic session, and fitters on a production bonus will not be very sympathetic if they are held up for too long. Also, if the equipment is ready for delivery to the customer, the financial department will be quite anxious to see it on its way. If all these questions can be answered satisfactorily, then photographs may be suitable, but normally line drawings will be more effective and economic.

The job and its tools

This section describes personal attributes and criteria for a career in information graphics. Its purpose is to help authors who are wishing to enter information graphics as a career and to make the uninitiated aware of just what is involved in the preparation of illustrations for publication.

It is normal for the illustrator to work in close liaison with the author of the documentation requiring graphic support, to ascertain the style of graphic and its content, before gathering the facts needed to prepare the final agreed images.

Where to start?

The professional illustrator requires the ability and flexibility to absorb details of a wide variety of technologies as well as the traditional competence in being able to produce the range of rendering techniques already described. They also need comprehensive communication skills with which to gather all the relevant information that is required to complete the briefs correctly and efficiently.

Whatever the equipment, system or process that is to be illustrated, it must have been designed and manufactured by someone, and this is the starting point for gathering data. Ideally this should be commenced at the earliest possible stage in the project. The illustrator must be prepared to seek out all the necessary available information such as drawings, parts lists and photographs, and interview anyone who has first-hand knowledge of the equipment, its assembly and functionality. They need to see the equipment, sketch details and take photographs. Careful study should be made of all the technical data and performance criteria in order to absorb as much information as possible, and clarification should be sought on anything that is not understood.

The illustrator needs to understand the manufacturer's drawing system so that they can readily find important information. They need to be aware of modifications, procedures and schedules to ensure the final illustrative rendering is current. The illustrator needs to be aware of any developments in the lifecycle of the project in order to ensure that their role in the project is completed efficiently and cost effectively.

How does it work?

Remember that the object of a technical illustration is to create a visual image of the functionality or construction of a product. It is essential, therefore, that the illustrator should understand the product too. Of course, no illustrator can be expected to be a technological expert on every subject, but sometimes the solution to understanding can be as simple as having a unit dismantled to clarify its function.

Use of camera

The illustrator does not have to be an expert photographer, but the camera is an excellent fast recorder of information and can be an invaluable aid. New digital high-resolution cameras offer the opportunity to obtain images that can be downloaded directly to computer and manipulated by the graphic software packages. Whenever there is an opportunity to record a stage during the build of the equipment, especially if modification or changes are possible, it is useful to take photographs. Such opportunities may rarely occur again. A photograph can also be used as a basis for an illustration, and even if the quality is not good an enlarged print can be traced. This is particularly helpful when manufacturer's drawings are not available and consequently cannot be used to construct a drawing.

Traditional illustration production

The traditional method of preparing illustration artwork for publication required the use of pencils, pen and ink, rules, curves and lettering stencils to render the image on a board and or overlay film.

The board on which such line drawings are produced has a smooth white surface ideal for taking a clean black ink line applied by drawing pen, ruling pen or brush. Line board is very durable material and can withstand erasing and even scraping with a razor blade for modification. The method requires a pencil drawing on tracing or drawing which is ultimately transferred onto the final line-board by tracing or 'pushing-through'. The drawback of this method is that it is labour intensive. In effect, each illustration is drawn three times - once on the tracing paper, again when it is pushed through on to the line board and finally when it is inked in. An alternative illustration method uses a translucent drafting film laid over the pencil drawing so that the image is then inked straight on to the film. This produces an original that can be dyeline copied for checking before being reproduced for printing.

New tools and techniques

The use of computer technology with graphic software packages now replaces the traditional tools for graphic production. These packages have developed at a tremendous pace over recent years, and their programs have much to offer the skilled illustrator. The scope of the computer systems are determined by the scale of the projects being worked upon. Programs are available to operate on individual PC installations or adapted and included on large multi workstation networks directly linked to publishing systems. The choice is enormous and each offers a high degree of consistency in the production of finished images. It is interesting to note that the function menus of these programs utilise the traditional nomenclature when creating computer generated artwork.

The selection of hardware is also largely governed by the scale of the publications project and the interfaces required with the other design, manufacturing and documentary aspects and scope of the projects requiring graphics. Large companies (such as automotive, aerospace, avionics and shipbuilding companies) who are preparing documents to exacting standards and specifications, and are employing numerous staff, find economy using the larger workstations which are fully utilised everyday. For the versatile small scale studio or freelance graphics producer economy and efficiency can be obtained from individual PC based hardware.

Both environments require the illustrator to become computer literate, but still require the traditional and aesthetic skills of vision and drawing in order to the support written documentation. The vision and the aims remain the same, it's just the tools that have changed.

Graphic software, like the hardware is largely governed by scale, whether it is used on a workstation where it has direct access to CAD information files and the documentation and publishing systems, or a compact project that can be completed on an individual PC.

The following is a typical list of computer aided illustration programs, a number of which need to be mastered for progression in information graphics as a career.

Technical illustration in mono and colour for on-line and conventional publishing use

- ITEDO Isodraw graphics - www.itedo.com
- InterCAP technical graphics - www.micrografx
- Adobe Illustrator - www.adobe.com
- CorelDraw - www.corel.com
- Macromedia FreeHand - www.macromedia.com

Retouching and photo-manipulation programs

- Adobe PhotoShop – www.adobe.com

Three-dimensional programs

- CorelDream – www.corel.com
- Adobe Dimensions – www.adobe.com
- StrataStudio Pro – www.strata3d.com
- Infini-D – www.metacreations.com
- Extreme3-d – www.macromedia.com

Whither illustration?

The art of illustration is dynamic. This places a response of adaptability and flexibility on those choosing it as a profession in order to meet the demands of publishers. The ability to visualise a complete illustration from the outset and plan the position of individual components on the page to best advantage is just as important as familiarity with each new software development.

It brings with it a challenge to build on the traditionally required skills by continuously exploring and absorbing the best of new technology, as many new skills must be learned.

The future for those in information graphics is greater than ever, in this age of electronic documentation. On-line publishing brings with it requirements for animation, digital presentations, web design, promotional and corporate publications, training aids, the creation of virtual environments and many other creative solutions.

Glossary of terms

animation	The production of a close series of illustrations depicting the relationship of two or more components regarding their dynamics one to the other, which when projected in a movie sequence give the impression of real-time movement.
annotation	Direct: The name or description of an item or an illustration, with a leader line pointing directly to the item. Indirect: A number pointing to each item on an illustration which is accompanied by a key identifying the items.
artwork	Originally this term referred to any hand lettering or illustration that was to be to be copied by the process camera for block-making, as opposed to type characters which were taken from the printer's case. With litho-printing it refers to all material that has to be camera-copied for plate-making.
bezier curves	A curve defined mathematically by four control points. These path between two points and are the means of producing smooth curved lines on a pixel based screen. They are used in conjunction with vector graphics.

bleed	When a printed colour is printed to the trimmed edge of the paper, the image is normally produced so that it slightly overlaps the size of final printed page. It is then printed on oversize paper and trimmed down to the final size required. The amount it is printed oversize is called bleed.
CAD	Computer Aided Design which normally works in three dimensions, and whose files can be accessed by some graphic programs to create illustrations. CAD programs are capable of representing 3-D images which can be rotated on screen, but whose pictorial resolution is usually insufficient for publishing purposes.
camera-ready	Text and illustrations prepared for the process camera, exactly as they will be reproduced when printed.
continuous tone	The tonal range from light to dark or white to black of a photograph illustration such as a pencil drawing, an airbrush drawing or a wash drawing.
cut-out half-tone	A half-tone photograph or illustration with the background removed.
digital colour	(RGB - Red Green Blue) The basic of additive primary colours used to create all other colours when transmitted light is used i.e. work for on-line display.
duo-tone	A method of double screening a monochrome picture so that it can be printed in either two colours or two tones of the same colour.

.eps	Encapsulated post script graphic file description.
frequency (lpi)	Lines per inch.
half-tone	A print of a continuous tone photograph or illustration reproduced using a series of dots varying in intensity to match the tonal range of the original.
inset illustration	An illustration set into a page of text or other illustration.
.jpeg	Graphic file compression.
landscape	An illustration or book, designed to be viewed with the longer side horizontal.
line drawing	An illustration that consists of only black lines or mechanical tints on white background, or the reverse, and therefore does not need screening.
lower case (lc)	The small letters such as a, b, c, etc., so-called because they were always kept in the lower box, or case, of a typeface.
upper case (CAPS)	Capital letters of a typeface such as A, B, C, etc., kept in upper box, or case, of a typeface.
mark-up	The preparation of a page of text indicating to the printer the size of type, spacing, margins, etc. required.
master	Original artwork from which copies can be made.

moiré	An optical effect which can appear on colour half-tone images if the screen angles are incorrect or out of register.
mono	A single colour illustration.
.mpeg	Movie compression file.
overlay	Part of an illustration drawn on transparent material, so that it can be treated separately from the rest of the illustration. The material used should be stable (not affected by atmospheric conditions) and keyed to the illustration by registration marks. For example, an illustration that is not annotated could have several overlays for the annotations, each in a different language.
SGML	Standardised general mark-up language.
html	Hypertext mark up language.
Pantone	An international colour matching system.
.pdf	Portable document format, which incorporates graphic text and font information. Can be read cross-platform.
portrait	An illustration or book, designed to be viewed with the longer side vertical.
pixel	A single dot on a computer display. Templates and bitmaps are collections of pixels.
process colours	(CMYK - Cyan, Magenta, Yellow and Black). The term used for the preparation by the camera of material for printing by the four colour process.

registration marks	Marks, usually crosses, used on separate pieces of artwork that have to be keyed together to make one multiple layer fit over another when printed.
resolution (dpi)	Number of dots per inch displayed on-screen or printed on an output device.
.sea	Self extracting archive file designation.
screen angles	The angles of the four screens CMYK layed over a continuous colour image in order to render it as a half tone for printing production.
squared-up half-tone	A half-tone photograph or illustration with the background made square or rectangular to the required size.
tiff	Tagged information file format used for graphic information.
vector graphic	High resolution computer generated illustration files which maintain their quality on-line or published even when resized.

Acknowledgement

The graphic illustrations in this article are courtesy of Dunlop Aviation Ltd.

8
Indexing

Richard Raper *has been an indexer since the early 1960s and is now Managing Editor of Indexing Specialists, a leading indexing firm located in Hove, East Sussex. He gained* companion grade *membership of the ISTC in 1993 and for several years presented seminars on indexing to members.*

This chapter outlines why indexes are necessary as retrieval tools for both paper and electronic-based information. Automatic indexing, computer-assisted indexing and indexing with word processor and dedicated software systems are compared and their relative merits considered. Indexes for electronic products are briefly discussed with reference to their application for hypertext linkages. The chapter continues by exploring the theoretical background of indexing and the literature available. It concludes with a mention of ISO 999, the international standard on indexing. The Society of Indexers, with its associated groups around the world, is acknowledged for its key role in creating indexing awareness during the past forty years. Selecting the right person to index a work is discussed, with advice on commissioning a professional indexer. The chapter concludes by offering a procedure for compiling an index that includes an outline of planning hints on choosing entries or terms, dos and don'ts, style of presentation and editing requirements.

Introduction

The explosive increase in supply of information in all its forms demands increasingly efficient systems for retrieving its contents. Growth in information emphasises the need for retrieval instruments, among which indexes can play an important role.

The sheer volume of text information now available creates an environment where the time available for reading it is under constant pressure. Hence, much information is relegated for reference use only. Savings in valuable search time are made possible for paper products with the help of effective indexes. Although many manuals on technical matters are well indexed, there are still a large number that receive only a cursory product, if there is one at all.

Making information available by means of a comprehensive list of contents is another line of approach, but its sequential arrangement usually makes a search more time consuming than with an alphabetically arranged index. Also a list of contents cannot, for reasons of space, provide ready access to more detailed content within chapters that does not necessarily involve the subject of the chapter. A comprehensive subject index also offers additional information, such as cross-references that support the possible need by users for guidance on related topics.

Search engines for locating electronic information, in disks, CD-ROMs and the internet, enable retrieval of much key information, but only if the search question matches the term sought. A searcher may not know the precise term describing the information required. Consequently, additional time is required in finding the information – if it is found at all. This is the sort of situation where an effective electronic index could offer extra benefits similar to those available for paper products.

The procedure for electronic index preparation is different for the reason that information presented in electronic versions (except for Adobe PDF files) is not usually identifiable by visible paragraph or page numbers, but by topic or subject content. Hypertext links are required that involve the use of tags (invisible to the user) embedded in the text. Such tags are cross-linked to entries in a separately presented index. Procedures for preparing indexes in this way are available, but the methodology for producing them is in urgent need of streamlining for reduction in the cost of their preparation.

Returning to paper based indexing, how is its effectiveness measured? Good indexes to publications should serve a relatively transparent function by supplying information for users with minimum effort. As an art form the presentation of an index is also important for giving the benefits of workability and simplicity. Indeed, its complexity is hidden by a seemingly simple arrangement of its contents. But users do not have to know this. Many an author compiling an index to their own work for the first time will discover that the process of creating, or compiling it, requires employment of different skills from writing so as to properly serve the needs of the readers of the work. So, how should technical authors in the ISTC proceed with compiling their own indexes?

This chapter aims to go some way to answering this question by offering a general procedure for preparation of an index, but before that, several alternative methods are explored in some detail. The choices or options include use of automatic indexing software, software that is already part of word processor packages and dedicated software packages for computer-aided indexes. The traditional way of producing indexes with index cards or small slips of paper is still used occasionally, but in most situations is superseded by computer-aided methods.

Automatic indexing software

Such software is used for simple indexes such as basic parts manuals, but it has its limitations. Hazel Bell discussed it in some detail in a recent paper that explores its abuse.[1] Every few years, it seems that another new software product is released with claims that it solves the problem of indexing with minimum effort. 'Just press a few keys on the computer and the programme does the rest!' Sadly this is not so, for several reasons.

Indexing is about the meaning of words and technical terms. Interaction between the relative location of adjoining words can offer different meanings compared to words that are isolated from the text. For example in a book about computers, it might identify the term 'computers' and index it every time it occurs in the text, but the index cannot identify their significance as the selection is automatic and not by meaning. A serious flaw occurs with automatic indexing because of the difficulty of coping with the mixed structure of spoken language. Because of its inconsistency, language is not easily amenable to the treatment of automatic indexing, because computers cannot truly identify terms for their meaning with certainty. Perhaps cognitive software may make it possible, but in my opinion it is not likely in the immediate future.

Word processor and desk top publishing software

This procedure is popular for manuals requiring frequent updating for some industrial machinery and software products. It is particularly useful for a digital printing process that prints and packages a complete manual in one pass, including the index. The method is not always successful if the software for printing is different to that used with preparation of the text, because the pagination sequence is altered with consequent incorrect page numbering in the index. The software may be major word-processing packages such as Microsoft Word, WordPerfect, Lotus or DTP packages such as Ventura, Framemaker and Interleaf. Two main methods are available: creation of concordances and tagging procedures.

Concordances create entries for every matched occurrence in the text of words/terms from a prepared list. Subsequent scanning of the text for the occurrence of the terms may produce an index that appears adequate until a check of the data demonstrates that many of the items selected do not always make sense. The effect is to increase the time required in order to verify that every entry is relevant to the contents of the text – this can take many hours rather than the minutes originally expected when generating the index.

The second route is to select significant terms or phrases in the text by means of index tags inserted into the text but transparent to its readers. After completion, the index entries are automatically extracted and arranged in alphabetical order and linked with the page numbers where they occur. Incidentally, it is also necessary to thoroughly edit the index and check each entry to ensure consistency and accuracy. The overall time taken doing this may be as much if not more than it would be if it were compiled manually using well established computer-aided procedures.

Computer-aided indexing

Dedicated indexing software is used in various forms, some specialised for particular situations. The most successful packages worldwide are Macrex, Cindex and SKY Index but there are others. They allow the human indexer to manually key in the necessary index entries directly from page proofs. They also facilitate the process with macros, automatic sorting of the entries and arrangement of acceptable page number or paragraph number locators. They also provide diverse printing options to produce any particular index style. The human-computer interface of this type of software is probably the most successful combination for production of good indexes.

Markup languages

The use of markup languages has been devised to supply standard procedures for structuring documents through to their publication in a variety of formats. It can involve use of SGML (Standard Generalised Markup Language) or a variant. Supported by international recognition (ISO standards), SGML makes possible the conversion of data for alternative publication with paper, CD-ROM and internet products. For the web, HTML (Hypertext Markup Language), which is compatible with SGML, is currently the most commonly used system for presenting data. Indexes can be located as tags in the text but, compared with traditional indexing processes for paper products, they can cost more. This is because extra coding is required. Metatags and search engines used for locating information are not indexing in the traditional sense with paper products, but are focused for electronic applications. More information on metatag initiatives can be accessed at http://www.publishers.org/mici.htm. XML (Extensible Markup Language) is being developed for the next generation of instruments for document structure control and data interchange on the web.

The principles of indexing

The rest of this paper aims to serve as a general guide for technical authors wishing to compile their own indexes. A background on literature about indexing is explored briefly, followed by advice on the steps that may be taken into account for compiling an index, such as planning, creating or writing it and editing of the product.

Background

Hazel Bell gives an account of the history of indexing and the development of modern professional societies during the past forty years in a series of papers.[2] Indexing societies are now functioning successfully in Australia, Britain, Canada, China and USA and more recently in Southern Africa. This interest has led to development of a profession of indexers to the extent that their members are able to support technical authors, if required, with indexes on any subject, or can serve in a consultancy role as required. The current books on indexing include *Indexing Books* by Nancy Mulvany[3] and *Indexing from A to Z* by Hans H. Wellisch[4]. Professional articles on techniques of indexing are regularly published in *The Indexer,* the professional journal of several Societies of Indexers in the UK and overseas. A book for indexers in the UK by Pat Booth is expected soon. In addition the *Anglo-American Cataloguing Rules*[5] is another resource that is especially valuable for defining alphabetical arrangement of names in different countries and their languages.

Although several countries have produced their own standards for indexing for many years, the international standard ISO 999[6], with examples, is now the world standard. It supersedes the British Standard BS 3200 : 1987 on indexing and BS 1749 : 1989 on sorting. It concludes with a model index compiled by Janet Shuter[7].

Some authors might find that organisations hiring them for a technical publication are not quite convinced of the benefits offered by a good index. Raper[8] discusses in some detail such benefits, and lists proposed criteria for measuring the qualities of a good index. If authors decide to index their own work, it is strongly recommended that sufficient time is allowed for index generation. This advice applies with both word processing software and dedicated indexing software. Whatever the method of preparation, enough time should be allocated for careful checking of the index for accuracy, consistency and useable style of presentation.

Authors or editors who prefer the services of a professional indexer can obtain guidance on commissioning an index from a helpful pamphlet available from the Society of Indexers.[9] Alternatively, they can contact the Society for a shortlist of its members selected for their subject specialism. The several steps that make for successful preparation of an index involve planning, indexing action, checking the product, sorting, proofing, editing and delivery in the form required for publication.

Planning

Planning plays the essential role of ensuring that the time spent preparing the index is effective. The table below seeks answers to the most important questions on specification of requirements.

Key questions on index planning

1. Number of pages of the text
2. Number of pages available for the index
3. Number of index columns per page
4. Number of lines per column
5. Line width in terms of average maximum number of characters
6. The calculated number of lines in total, allowing extra space for the heading and any notes.
7. Estimated number of page references per line
8. Expected number of actual index entries per page
9. Style of presentation or layout of the index for ease of use.

Items 1 to 5 depend on page size, the point size of the chosen font or typeface plus its body size which defines the density of lines to a given page length. If the number of pages available for the index is known, then a specification for the total number of lines for the index is calculated and thereafter the number of paginated entries available. If still uncertain, then index a few pages experimentally and from this information determine the number of index entries per page for an effective and efficient index. If the two numbers agree then go ahead with indexing the rest. If not, either negotiate with your editor/author for more pages or rethink the number of entries that will suffice.

Popular layout styles are 'headers' with indented subentries or even sub-subentries, or run-on or paragraph style when subentries are continuous. Although this latter approach is attractive from the production point of view as it saves valuable space, especially for larger indexes, the subentry style is easier for quick eye scanning by its users.

Selecting index entries

After reading the material awaiting its index, choosing topics for entries is the next step.

The publication may require separate author and subject indexes or a combined index for both. Whether or not to have several indexes is a problem that bothers those who think that a larger index is more difficult to use. Not true, and the best way of determining an answer to this question is a brief survey of other publications. Incidentally, ISO 999 prefers one index but allows for different options.

Listing authors should follow the literary style used in the work being indexed. Some quote every author, others the first author of a multiple author paper. Some use first names, others initials only with or without full stops and some adopt the convention of running initials together; for example, L.C. or L. C. become LC.

The aim of a good subject index, as a retrieval instrument, is to provide an objective and balanced analysis of the subject sufficient to make all key material in the text accessible – people, places and things. Adjectives, verbs and adverbs should be avoided, because they can be misinterpreted by users and even become misleading. Nouns and present participles are more useful as they are direct and less likely to suggest undesirable bias. Compound words or phrases, popular for certain technical terminology, break all these rules, but such terms should be used as if nouns. The grammatical form of some words is mixed. For instance, 'green' as a prefix is an adjective and should be avoided, but if discussed as a colour element or 'noun-form' it becomes an entry in its own right.

Writing index entries

There is no standard procedure during the compiling process when making an index, since it varies from person to person according to their mental make-up and related skills. Pat Booth[10] presents six different approaches for compiling an index. Some write their index in a single computer file from the beginning of the book through to its conclusion. Others, including me, compile their index in a series of shorter files, checking the contents at each stage of file completion. The individual files are finally merged into their book form. At all times, the specification of the index, established during the planning stage, is put into practice. It is too easy to write index entries whenever it seems appropriate and find later that the index is one and a half times over its planned length, which might have been avoided if the specification had been followed. Substantial reduction of an oversize index is very time consuming and should be avoided, but some reduction is useful as it tightens up the presentation of the index without losing meaningful content.

On completion, it is essential to verify that the length of the index and page density of entries satisfy the specification, and adjust them as necessary. It is also important at every stage to proofread the index and check that all data is there. It is too easy to miss a file or piece of information without checking and re-checking it at every stage of its preparation. Finally the index is typeset for its publication. If possible, carefully check the proofs and eliminate inadvertent errors that are introduced during typesetting. Examples are: headings placed on the last line of a page before indented subentries, turnover lines not indented correctly, capitals that should be lower case and *vice versa* and, worst of all, garbled text.

References

1 Bell, Hazel. 'Perilous powers in authorial hands.' *The Indexer* 21 (3), 1999, 122-3.

2 Bell, Hazel. 'History of societies of indexers.' *The Indexer* 20 1997, 160-4, 212-15; 21, 1998, 33-6, 70-2; 1999, 134-5.

3 Mulvany, Nancy (1994). *Indexing Books.* University of Chicago Press, Chicago and London.

4 Wellisch, Hans H (1991). *Indexing from A to Z.* H. W. Wilson Company, Bronx, N.Y.

5 *Anglo-American Cataloguing Rules, Second Edition Revised.* Library Association, ALA, London (1998)

6 ISO 999: 1996(E*). Guidelines for the content, organization and presentation of indexes.* International Organization for Standardization, Geneva (1996).

7 Shuter, Janet (Ed.) *ISO 999 : 1996 (E)*

8 Raper, Richard. 'The need for book indexes, indexers, types of indexes and some search techniques.' *Aslib Proceedings* 42 (7/8), 1990, 207-12

9 Society of Indexers (1998). *LAST BUT NOT LEAST. A Guide for Commissioning Editors.* Society of Indexers, Sheffield.

10 Booth, Pat F. 'How we index: six ways to work.' *The Indexer* 20 (2), 1996, 89-92.

9
Document distribution

Pete Greenfield, FISTC *has worked in the Technical Publications industry for 26 years and was President of the ISTC between 1992 and 1994. He joined Abbey National in 1987 to set up the Business Communications area and was project manager for the on-line viewing project, which converted paper documentation to computer-based and installed a satellite distribution network to deliver information data to the Abbey National network. He is currently a member of the NVQ steering group which is looking at vocational qualifications for technical communicators with the Department for Education and Employment. He was responsible for the implementation of ISO9001 for Business Communications at Abbey National and is currently developing intranet to the retail division. Peter was awarded the Horace Hockley award in 1995 by the ISTC, and the 1998 Frank Chorley Prize by the City and Guilds Institute of London.*

Introduction

There are a number of ways to generate text, both on paper and on-line; indeed the majority of development and software is dedicated to this outcome. A problem for technical communicators has always been how to distribute this information to the user and how to ensure that it is maintained in the future. On-line and computer based documentation solves much of this problem, although these methods now present the additional problems of uniformity of viewers, desktops and protocols. The problems of finishing printed matter have moved into the technical problems of IT.

As standards emerge, such as HTML, XML and so on, the argument for methods of distribution can depend on a number of parameters. These will include number of users, geographic spread, networks in use, volume of information, concurrent usage required, volatility of information, critical nature of the information and so on. The arguments in broad terms boil down to whether to centralise the data or whether to distribute the data.

For situations where the user spread is not known and the critical nature of the information is low, a centralised development is usually the chosen course – for example, the intranet. For large companies with a large user spread, even the intranet may not be resilient enough and other solutions will be required. The communicator of technical information may find that they need to 'buck' trends to ensure that the user is delivered the service required. Items for consideration are likely to be the size of the library to be delivered, availability required, special software in use, special desktops in use, screen sizes available, amount of change, and search features required. The case for intranet or internet style communication is not always as straightforward as it may seem.

In the case of Abbey National, the requirement was for a secure, 'push information' system with unlimited capacity and 99% reliability. The chosen path was a satellite distribution system. This particular system has been designed and implemented for a corporate need, but the benefits and principles will be mirrored in other organisations and show that, although system integration and industry standards cannot be ignored, the business need can drive the solution rather than simply using fashionable solutions.

Business issues

Abbey National plc is one of the UK top six banks. The financial sector is currently going through extensive change and is subject to high profile legislation requiring compliance with both the Financial Services Act and the Consumer Credit Act. Documentation plays an extremely important part in ensuring that the bank remains compliant with these Acts.

The Business Communications department is responsible for the issue of all policy and procedural information to the Abbey National retail network – a readership of over 17,000 staff. The documents cover change in procedure and legislation, introduction of new products and systems and personnel issues.

In 1996 the volume of communications that branches received, the difficulty of absorbing the information, and the problem of finding the particular piece of information when required was beginning to become a barrier to the pursuit of business, with an impact on customer service and staff efficiency. The volume was, however, totally driven by business need.

It became necessary to radically rethink the communication process, in terms of both corporate communication strategy and technology used. An investigation into alternatives was carried out and recommendations made. The culmination was a £3.5M project that introduced on-line documentation updated via a satellite distribution system.

The project developed a return on investment of over 260% and laid the cultural foundations for the development of further on-line information and the eventual introduction of a company intranet. In environmental terms Business Communications has reduced its paper output from over three million A4 sheets per month to under 100,000 sheets per month.

As a technological spin off, the satellite distribution network is now used to distribute business television to field staff. It is also used to distribute software changes and upgrades to the field, an activity previously achieved by using the transaction network. The use of the satellite system has reduced the distribution of software from weeks to days.

The new system has brought real business benefit but, more importantly, has given Abbey National the ability to provide documentation and communication as a core function for all staff, rather than the 'necessary evil' that it tends to be in the majority of companies. While the system currently provides a 'push' process it does allow information to be provided at the desktop and it is a short step to provide for interactive communication, when the time is appropriate, now that the infrastructure is in place.

The issues that drove the implementation of this project were based on the effect that poor communications had on the business. Within Abbey National, customer loyalty and service is a high priority and anything that improves service is of great value. Poor communications affect the knowledge that staff have and this is seen in:

- Poor service
- Danger of non-conformity
- Poor image
- Staff time taken on update, which translates as fewer staff serving customers.

The requirements

Business Communications is responsible for the notification of new products, product changes, new systems, new procedures and changes in legislation. These are business critical documents and therefore it is essential to the business that the documents are easily accessible and that they are up to date. This especially applies to legislative documents. The mission statement is 'to provide timely, accessible and authoritative' information.

As a media for this type of documentation, paper is costly, difficult to distribute, difficult to access and users cannot be relied upon to update the information. For some time it had been a wish to remove paper and several systems had been tried. As the library comprised 60,000 plus A4 pages and approximately 60% of that total volume changed each year the problem had always been: How do you distribute that volume of information reliably, quickly, flexibly and cheaply?

The business demands that, as a department, Business Communications reacts to business initiatives quickly. The distribution channel had to be able to react to ever varying volumes and be permanently available. Also, disaster contingency was an essential part of the required package.

The financial industry is very competitive and, in terms of interest and new products, is reactive to the market place. The speed of new product introduction and the range and complexity of products being offered is increasing at an alarming rate. With the introduction of the Financial Services Act, regulation of the industry means that documentation and understanding of that documentation is more critical than ever. Especially critical is the maintenance and currency of any documentation.

System operation

The system is designed to distribute text and graphics data to each site, from a single central distribution, so that it can be searched and read from a viewer on each computer in the branch. Data is also sent back to the central site to provide an audit of receipt and to automatically generate a fault call on the receiving equipment if necessary.

Once the information is completed by Business Communications it is distributed via the satellite network and access given at the field terminals. The information is kept up to date centrally, and requires no user intervention. The information is accessed by a series of navigation tools and search facilities from the branch's local area network.

The distribution of information is split between field sites and Head Office sites. Field sites use a satellite distribution to deliver information to each local site. Here it is stored on the local file server and accessed by each workstation. This method is economical because of the number of sites served, and presents a disaster recovery solution as all data is held locally, providing unlimited backup to all sites.

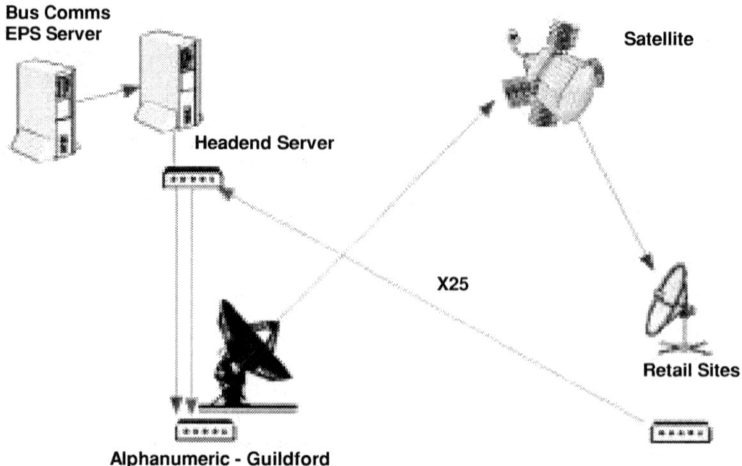

Satellite system

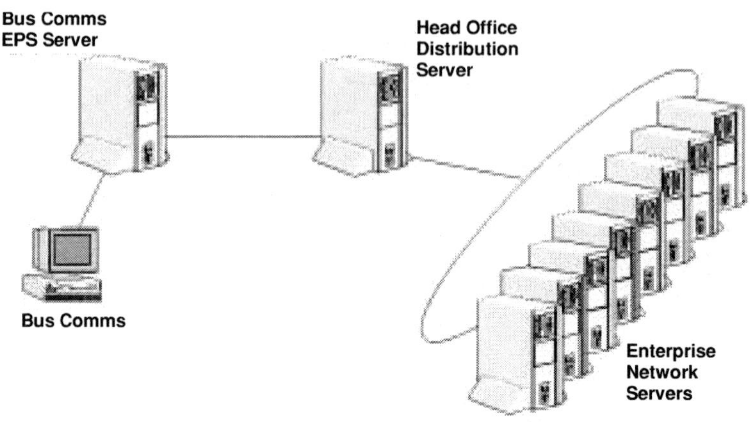

Head Office system

Business Communications

The data files are prepared by Business Communications in Abbey House, Central Milton Keynes and transferred to the Data Centre via the Company Enterprise network. From this point the Data Centre controls the distribution to the field.

Within Business Communications, authors using Microsoft Word, create information. A database, specially written in Business Communications, tracks the work and the versions, and is used to manage the workflow and provide management information. Once the author has completed the work the files are sent to the Publishing Section, using Rich Text Format (RTF).

The Publishing Section uses publishing software to prepare the work in Abbey National brand format, add all internal document management and add any illustrations necessary. A database is used in Publishing to provide compliant file names for the electronic press and ensure there is no duplication in file names or versions. The publishing software provides a source document which is managed and can be issued either on-line or on paper from the single source.

Once the source document is completed it is pressed to provide the on-line hypertext version and is compiled with any other changes that are to be issued. This type of production is essential for large volatile documents where the company has to be sure that the correct versions are available and needs to track, and sometimes make available, older versions.

Once the daily collection is completed it is sent via Wide Area Network to the Data Centre for distribution. In essence, Business Communications provides a directory and file structure that is duplicated throughout the Abbey National information network.

Data Centre

Distribution

The Data Centre production and off site backup provides the addressing facility and database audit trail of distribution. The system enables the operator to target the sites, from one to all sites.

The compressed data is transmitted to a satellite communication provider, via a Kilostream line (128 Kbs).

The satellite provider stores the data until a transmission slot becomes available – the contract dictates within four hours. In practice the timescale is considerably less. The data is transmitted to the uplink site for onward immediate transmission through the satellite at up to 6 Mbs.

Data is sent in one direction only for this particular solution.

An existing branch computer processes the data with an internal decoder card. The workstation passes the data straight through to the branch network server. There it is decompressed, decrypted and stored.

The server indicates, back across the standard branch network to the Data Centre system, that the files have been correctly received. The Data Centre also has a satellite broadcasting monitoring system to ensure that the satellite is transmitting correctly.

Should one or more sites not receive the full transmission, the Data Centre re-transmits to the missed sites. To aid recovery, large broadcasts are split up into small packages so that only the missed packages need be re-transmitted. In this way full disaster recovery is achieved as each branch has its own full set of files.

Business Communications sends large amounts of data to each site daily. Of course the output from Business Communications is totally dependent on business need and it is impossible to book in advance the bandwidth required. This was a major factor in deciding on satellite distribution as this was one of the only methods with spare capacity and speed to cope with this flexible need.

The channel is *not* allocated to Abbey National 24 hours per day, seven days a week. The satellite provider commits to transmit within four hours from receipt of the files. The system has proved to be cost effective based on the number of sites served.

ISDN lines have been connected to a few sites where it has proved to be impossible to mount dishes. This has been because of local leases or the type of building. The satellite supplier manages the ISDN distribution and Abbey National still only provides a single feed, making the service transparent.

Proof of concept pilot

Implementing such a system is a 'once only' chance so it is essential to test the concept as well as the technology. User input is essential to ensure that the library structure chosen is acceptable.

The benefits outlined in the Business Requirements Specification were proved by proof of concept pilot. The software was popular with staff as well as robust and reliable. However it was also proven that any future roll-out of the system needed to pay particular attention to the training of staff, as there is a culture change in using documentation on-line, compared to using the traditional paper-based documentation.

Scope of the pilot

Aims

The pilot was set up to act as proof of concept for on-line documentation and not as a working model. The aim was to prove the cost and efficiency savings outlined in the Business Requirements Specification, as well as to gain valuable user feedback on the system.

Pilot sample

Access to the agreed on-line viewing system was given to three branches, a mortgage service centre, a banking centre and a Head Office department for a period of 16 weeks.

Outline

The browser was provided for use in the pilot field sites via a number of stand-alone workstations, and in the Head Office department via the local area network. A selection of procedure manuals were provided and updated on a weekly basis.

On site training, quick reference guides and a help line were provided by Business Communications for all pilot site staff. During the pilot, staff were not allowed access to the paper versions of the manuals available on the system.

This was a deliberate decision that also proved to be necessary when the system was fully implemented. It was found that staff would always go back to what they were familiar with. To break this the system needs to provide a major advantage over any existing system. Although staff identified the major advantage of this system to be that they no longer needed to update manuals, for the majority the familiarity of paper needed to be broken. To do this it was necessary to remove paper. Later research showed that once staff were familiar with the system the majority preferred it to paper-based manuals, citing access as the key advantage.

Proof of benefits

Summary of tests undertaken

Towards the end of the pilot, a number of tests were carried out on members of staff from the field who took part in the pilot. The tests were designed to measure and compare the following between on-line and paper-based information:

- Speed of access to information required
- Accuracy of understanding
- Length of retention.

Speed of access

This test was designed to ascertain whether presenting documents on-line would quicken, have no effect on, or slow up the user in accessing information, compared to using paper-based documents.

The test was carried out by giving all candidates ten questions relating to their business area, five of which were completed using the on-line system and five using paper-based documents.

The time taken for each candidate to find the correct answer was recorded and correlated to reveal the following results:

- **Paper-based manuals**
  Out of a total of 50 questions asked, 49 questions were answered correctly at an average time of one minute 27 seconds per question.

- **On-line manuals**
  Out of a total of 50 questions asked, 45 questions were answered correctly at an average time of 31 seconds per question.

It can be seen that candidates were substantially quicker in finding the correct answer using the on-line system, compared to the paper-based system. On average, candidates using the on-line system found the answer in a third of the time of that needed to locate the answer in the paper-based documents.

Although the candidates using the paper-based manuals were slower than those who used the on-line system they were more successful in finding the correct answer in the allotted time of five minutes per question. This is illustrated by the fact that only one question remained unanswered when using the paper-based manuals, compared to five remaining unanswered when using the on-line system. Observations of the candidates using the on-line system revealed that certain candidates still had some training needs, which accounted for the questions remaining unanswered.

It is felt that with further training, coupled with more experience of using the system, the number of answered questions would be comparable, if not better, than those achieved using the paper-based documents.

In separate tests carried out with staff from Lending Development, who are Microsoft Windows proficient, the speed of access times were 15% quicker using the on-line system than those achieved in the field pilot sites. A larger percentage of questions was also answered correctly in the time allotted compared to the field staff. Therefore, further gains can be made on the speed of access if staff are already experienced users of Microsoft Windows.

Accuracy of understanding

This test was designed to ascertain whether presenting the same information on screen, rather than paper, enhanced, hindered or had little effect on the users' understanding of the subject.

The test was carried out by providing candidates with an exercise, on the on-line system, that gave them a set of instructions with diagrams on how to build either one of two types of paper aeroplane.

Candidates were observed when carrying out the exercise and the aeroplanes were graded, depending on how well the candidate had understood the instructions. These grades were then compared to those that were attained by different candidates who had previously completed the exercises using paper-based instructions.

The results of the tests showed that providing the instructions and diagrams on-line, marginally enhanced the candidates understanding of the procedure. This enhancement can be mainly attributed to the on-line document's use of coloured diagrams, compared to the black and white paper instructions, and the presentation of each step on its own, rather than part of a full A4 page.

Length of retention

This test was designed to ascertain whether presenting the same information on screen rather than paper, enhanced, hindered or had little effect on the user's ability to retain information.

The test was carried out by providing candidates with a mock Bulletin that they were asked to read. At a later stage, candidates were then asked questions relating to the Bulletin, without being able to refer to it.

The results of the test showed that candidates were marginally better at retaining information using the paper-based instructions. However, as the difference was marginal, it can be concluded that providing documents on-line would not cause significant problems in this area and, as reference to procedure had been proved to be quicker, the business operation would not be affected at this point.

Time spent using paper-based manuals at present

For a period of four weeks, four branches, a mortgage service centre and a banking centre were asked to keep a log of the time that they spent using paper-based manuals and Bulletins.

The logs revealed that, on average, each site spent one hour per week using manuals and Bulletins. If this figure is multiplied by the total number of branches, mortgage service centres and banking centres, and then calculated to give a yearly figure, then in the region of 9,880 hours are spent each year in the field, accessing paper-based manuals and Bulletins.

As demonstrated by the figures on speed of access, using on-line manuals and Bulletins reduces access time by 65%. Therefore, if the system were made available to all sites, in the region of 6422 hours would be saved. This could result in an equivalent cost saving of £96,000 per annum. (This assumes a cost to the company of £15 per hour.)

The speed of access tests did not measure the amount of time that it takes a staff member to actually go and find the appropriate manual to begin with, as these were already provided. Therefore, the time savings stated would be further enhanced if this were taken into account.

Time taken to file and distribute manual amendments and Bulletins

For a period of four weeks, four branches, a mortgage service centre and a banking centre were asked to keep a log of the time that they spent filing manual amendments and filing and distributing Bulletins.

The logs revealed that, on average, each site spent 35 minutes per week undertaking this task. If this figure is multiplied by the number of sites and then calculated to give a yearly figure, approximately 3725 hours are spent each year in filing manual amendments and distributing Bulletins.

If the on-line system were made available, considerable savings would be made, as field sites would not have to perform these tasks. This would result in an equivalent cost saving of *number of sites in implementation* x *minutes saved* x *£cost per hour of employee.*

Pilot constraints

A number of constraints affected the findings of the pilot. If these constraints had been removed there would have been no doubt that the benefits outlined previously would have been further enhanced. These might be typical limitations at the start of any implementation.

Staff not familiar with using Windows software

The majority of staff in the field pilot sites had never used Microsoft Windows software, or used a mouse.

The technical experience and IT knowledge of the user is an important factor to the success of on-line material. The author cannot assume the level of competence of an adult reader in navigating documents at the initial stages of using the system.

Documents not specifically designed for on-line use

The manuals and Bulletins provided as part of the on-line pilot were exact copies of the paper based versions produced by Business Communications. These documents were designed to be read on paper and not on a computer screen. For example, existing paper communications are printed on A4 sized paper, while the amount of information that can be displayed in the same format on a computer screen is about half of this size. This led to the system not being used as efficiently by the user as it could be. Taking the example above, the user had to use the 'page down' facility for each current page displayed on the on-line system, to reveal the second half of the A4 sized document.

For rollout, the legacy documents were converted to screen size to ensure the integrity of each page.

Help line analysis

Breakdown of calls received

During the period of the pilot, a total of 14 calls were received:

- Eight calls related to errors contained within the manuals, either spelling or mistakes in composition
- Two calls were requesting information on how to use the system
- Three calls were in connection with network access problems experienced by Lending Development
- One call related to a known minor software error.

Help line calls should not be omitted from any financial justification. Problems with the use of paper documents cannot be easily measured but audit trails on help calls and system usage can provide vital feedback for improvement and system justification.

Other findings

Culture change

It was observed during the pilot that a number of staff located information within the on-line documents by using the contents pages and then scanning the various pages, rather than using the more efficient search facility. The staff were in effect using the on-line documentation in the same way that they would use paper-based linear documents. As staff were initially not proficient in navigating around the documents in this manner, some difficulties were incurred in locating the desired information. Most of these difficulties would have been avoided if the search facility had been used to locate the information.

Similarly, some staff who used the search facility when trying to locate a particular piece of information, would actually search for the subject heading rather than the piece of information itself.

Staff needed to be re-educated in the documents that are provided on-line.

It was generally observed that there is a difference between using linear style publications and the random non-linear type of document that an on-line presentation tends to generate. Some of this difference lies with the author. The tools used for on-line documents encourage the author to build the document in a certain way, but at the same time this does not encourage the author to consider the user, a foundation of the technical author's responsibility. In on-line document construction, a new set of considerations come into play. For instance, in on-line documents the user cannot physically estimate the size of blocks of information, nor the structure of the content. Our user pilot showed clearly that users had a pre-set view of the document and how best to navigate around it. This was not always correct, and if the document itself did not assist the reader in navigation, its use was not efficient.

Communicators need to ensure that the technology does not drive the document, but that its use is the driver of the design. They must consider the structure, signposting information and its relative information, and the navigation of the document, and ensure that these items are consistent across the document set. They must also ensure that the terminology is consistent, not just to avoid ambiguity, but to ensure the search facility can find results reliably and consistently.

This brings additional training needs for authors. In training the user of the information, we found that the tendency was to train in the use of the software, whereas hindsight shows that training should be centred on the use of the document set. Browsers are relatively easy to use and are becoming universal in their use. Document sets need to fix in the mind of the user, and the signposting needs to be understood. It must be remembered that technical documentation will not be 'surfed' and that the user will quickly lose confidence and become frustrated if they cannot find the relevant piece of information; this is especially true if they are using the documentation in front of the customer.

These points are salient to any on-line system. The training required to use the actual browser and system was minimal, but the training to change culture and to understand document layout and structure had to be achieved over a longer period of time by reinforcement and by delivering what people needed to do their job. In effect, people will not change unless there is an advantage to them.

Proposed benefits

Benefits have been identified as:

- Central control of all information, especially compliance information
- Improved distribution of information
- Improved availability
- Improved accessibility
- Improved customer perception
- Faster retrieval of answers to problems and queries
- Reduced reliance on local 'oracles'
- Reduced calls to Head Office help lines
- Eradication of local information
- Eradication of personnel time to update information at local level

- Improved dissemination of information at local level
- Reduced reliance on local person to distribute important information
- Guarantee of local information being up to date
- Confidence that the most current level of information is referred to
- Assurance of legal and compliance reference to information
- Implementation of an information system that can easily grow with the company.

Other intangible benefits identified

- All users have manuals at their workplace
- All users have the same, up-to-date information
- Quicker, more effective publication process - sign off today, publish overnight, use tomorrow
- Daily updates possible
- Guaranteed delivery of documentation
- Potential for:
    - Greater procedural compliance
    - Greater accuracy
    - Reduced errors and queries
    - Reduced diversionary effort and time
    - Enhanced customer service
    - Staff satisfaction at doing a good job better.

Quantitative yearly benefits

- Reduction in employee hours for manual updating and filing
- Savings in document printing costs
- Reduction in the time spent searching documents.

Bottom line

- Return on investment 268%
- Payback 0.93

Summary

The system is now in its third year of use and the benefits have been driven from the installation. In company terms, the benefits have been fully derived and further use of the system is planned well into the future. The tangible benefits, which were identified early on, have been met as:

- Central control of information with audit trail provided
- Guaranteed update
- Update time saved by staff
- Information available to all, all of the time
- Improved customer service.

The rules, tests and principles applied to this project could be applied to any other similar system, whether it be bespoke or not, and should be applied to a major implementation of on-line documents.

10
How to grow a web site

Clyde Hatter, MISTC *is an Internet project manager specialising in content development and web site usability issues. Clyde has a background in technical communication and has served as an ISTC Council member.*

Introduction

Audience

This chapter is aimed at technical communicators who are involved in the area of web site development. The advice will be of most use to communicators who are moving from small-scale sites onto more ambitious projects. The article assumes familiarity with the basics of HTML mark-up and site assembly.

Purpose of this chapter

This chapter offers advice about how you can design your site so that it will grow over time, rather than fossilising soon after its initial launch. The advice in the article is general in nature and can be applied to many small-scale web sites that have been created with future expansion in mind.

Relevance of the topic to technical communicators

Web site development offers a major opportunity for technical communicators. Our core skill set is well suited to the all-round challenges of web development. In addition to originating new content, we can bring to bear the interpersonal, organisational and technical skills we possess to the lucrative and challenging area of web site design and management. By rebranding ourselves where necessary as 'webmasters', 'web editors', 'web site designers' or even 'internet publishing consultants' we can increase our worth in the employment market and also open up our careers in a multitude of new and exciting ways. Editing and illustration skills, the ability to structure and chunk information, interviewing skills, project management experience, knowledge of automated publishing systems as well as experience with hypertext systems such as Windows Help make technical communicators ideally positioned to become central players in the Internet world.

However if we, as communicators, are to make this leap we need to be aware that, in addition to extending our existing skills, web development will present us with new organisational and technical challenges. This chapter presents a series of tips intended to help you better meet these challenges and to help you avoid falling into some of the more common traps that await new web developers.

Establish your priorities

To develop a successful web site, you must balance how you divide your time and resources between those activities connected with content production and those connected with site design and maintenance. *Content production* includes the sourcing, commissioning and creation of new material such as text articles, news stories, training courses, diagrams and illustrations. *Site design and maintenance* includes the time spent marking-up text, producing navigational elements such as icons and buttons, maintaining scripts and applets and the activities associated with incorporating new content into the site's structure. If you get this balance wrong the end result may be a technically leading-edge site which is visually rich but which does not contain the content that your audience seeks. Or, conversely, you may have a site which is content-rich but which is poorly structured and visually unattractive.

Each time you incorporate a new element into your site – such as a hit counter or a navigation graphic – be aware of the long-term maintenance costs which the new element will entail and ask yourself 'is this really worth it?' Having an elaborate graphic on every page may add visual interest to the mix, but it may take up a slice of your budget that would be better spent on other content. Try to avoid adding elements which do not deliver significant added value to users of the site but which will result in heavy long-term maintenance costs.

Ensure a supply of content

The biggest challenge for you, as a web site developer, will be to secure a regular supply of high quality content for your site. Often web sites are launched in a blaze of excitement, only to fossilise as the supply of material dries up. The most common reason for this is potential information suppliers have no *compelling* reason to pass the information through to webmasters. A further reason is that information is forwarded in an unsuitable form, and the amount of processing necessary to incorporate the information into the site makes it prohibitively expensive. To avoid these problems you can use the techniques detailed below.

Create a development plan for the site

Development of a web site does not stop at its launch. If you want your site to continue to grow and develop, you must have a plan in place for its post-launch development and expansion. Since the plan will, almost certainly, draw upon human resources outside your direct control you must have the plan in place and agreed well in advance of its implementation date.

Introduce content in phases

You may find that the initial launch of the web site provides a stimulus for business units to provide a great deal of content. Following this initial burst of enthusiasm, you might, however, find it difficult to get follow-up material. This can result in the launch of an extensive and initially promising site, but one that fails to live up to its initial promise. To avoid this, consider introducing content on a phased basis. This gives your audience a reason to revisit the site and, if managed correctly can help to develop a spirit of competition among the business units which are supplying you with information – 'Have you seen that so-and-so has something up on the web now? Better update our own section!'

Establish formal agreements with your content suppliers

A web site can only grow if it receives a continuous supply of content, such as product information, news items and feature articles. You can only integrate multiple sources of information if you actually receive the information from these sources. To ensure that this will be forthcoming you need to have a clear idea of who your content suppliers are and how they will supply you with material. Content suppliers may include the business units in your organisation and may include external sources.

If possible you should have formal agreements with suppliers, detailing what they will provide, in what form and with what degree of regularity. Establish that they have the time, budget and resources to do as they say. Beware content suppliers who agree to provide material if and when they can. If a supplier of mission-critical content cannot commit to supplying that content, you must make the sponsor or owner of the site aware of the fact.

Educate and train content providers

To reduce the amount of conversion work required, work closely with your content suppliers. Can their authoring processes be modified to create web-ready output? Provide consultancy, training and support in return for commitments from suppliers that they will provide you with content that meets specific standards. Such a process benefits both parties. It reduces the your workload by providing you with materials in a web-ready form; it benefits content suppliers by adding to their skill set and by facilitating a rapid turnaround of their content.

Get management on your side

You will never secure a steady supply of content until the production of such content becomes part of the content supplier's business objectives. So long as content production remains voluntary, you will always be at the mercy of the urgent project or competing deadline. The only real way to address this problem is to get management to commit resources to Web content production and to ensure that targets are set for the supply of information to the Web site. However, this is unlikely to happen unless management understands that such targets are essential for the site's success. Make clear from the start the resource commitments that are required.

In order to get management on your side consider preparing an executive briefing on the technical and resourcing issues raised by web site development. The more clearly you present and communicate these issues, the less likely you will be judged against unrealistic expectations. The internet knowledge of many senior managers is acquired through browsing top e-commerce and business sites – usually without any thought to the cost or the infrastructure necessary to produce such a site. Unless you can set expectations correctly you will not be compared with your competitors, but with sites such as Amazon, the BBC, Bloomberg, eBay or Microsoft. When setting expectations, communicate clearly the areas of the site which you can develop quickly and for little cost, and those areas (for example, the processing of on-line payments) which have more significant cost implications.

Keep your design simple

Avoid complex text mark-up

Among the ways of adding visual interest to a web site are page-layout techniques derived from magazine and book publishing. These techniques use combinations of display fonts, text spacing, colour and alignment to emphasise the structure of the information and to hold the reader's attention. However, HTML mark-up does not make it easy to achieve these effects in an elegant fashion. Many web pages contain complex sequences of text formatting commands combined, in some cases, with the insertion of bitmapped graphics to achieve effects such as three-dimensional text and graduated fills. They are costly to implement in the first place and, if the style of your site changes, or if the information is to be presented in a different context, will need to be extracted and replaced by a new set of commands.

You can, however, use techniques which, while avoiding the creation of complex mark-up, still allow you to create stylish, visually attractive web sites. Among the techniques are:

- Control of text formatting using style sheets
- Display of text within a demarcated area. Use of the surrounding elements to provide graphical impact whilst using a simple formatting scheme for the text itself

- Use of a portable document format such as Adobe Acrobat PDF to present large volumes of complex formatted information. Although not as immediate as HTML, browser support for such formats is growing in sophistication. If, for example, you have a specialist 80-page report to present, complete with equations, tables and line-art, ask yourself whether it is really worth the cost of converting this to HTML. Publications intended for mass-audiences may justify conversion costs; those intended for small groups of specialists (particularly computer-literate specialists) probably do not.

Make use of style sheets

Style sheets are familiar to most technical communicators who have worked with DTP packages. Style sheets simplify text mark-up by associating a group of formatting commands with a single mark-up code. On the web this is achieved by using a CSS (cascading style sheet) file to link formatting attributes with HTML tags. A page can then be linked to a style sheet by a single reference; the style sheet then controls the appearance of all the text on the page. However, beware of the fact that different browsers interpret style sheets in different ways. For this reason you will need to check the appearance of your style sheet formatted pages on a range of browser types. The ISTC site makes use of style sheets to control text formatting.

Go easy on your links

It is not good practice to include a high density of external hyperlinks within blocks of text. Too many links within a section interrupt the prose flow unacceptably. External links, in addition to directing traffic away from your site, soon get out of date and therefore need to be validated on a periodic basis. Therefore, be restrained in the number of links that you add. Only include links that have a high relevance and, in addition, consider grouping the links together at the beginning or end of a page or section. This will improve the clarity of your site and will make the links easier to identify for validation purposes.

Use plug-ins and applets only if essential

If you produce content for your site that can only be viewed by installing a browser plug-in or by invoking an applet, be prepared to deal with enquiries from end-users who are having difficulty locating, installing or using the plug-in. The decision to include a plug-in should not be taken lightly. Not only will you need to assess the implications for your user support structures (help desk, and so on) but you will also need to extend your test plan to assess the reliability and compatibility of the plug-in. Does it work across the whole range of browsers that your target audience possess? Does it affect performance? Will it be affected by user's browser security settings? Whilst, in some cases, plug-ins are the only way to include essential functionality, often they are included gratuitously to give impression of a cutting-edge web site. If you wish to minimise a site's maintenance overhead you should investigate whether or not there are more straightforward alternatives.

Prepare page templates

Consider making all of your content pages follow a standard layout. Often this will consist of a standard header and footer containing graphical and navigational elements that sandwich the page's main content. The advantage of this kind of design is that you can prepare a template for the page that contains all the standard elements. You can then prepare the main content separately and just paste it into the template. Most web page development environments support this kind of approach.

Test your site

Develop a testing policy

The user's hardware and software environment can play a large part in determining the quality of the browsing experience. Different browsers, different security settings, different screen resolutions and different modem capabilities can all exert a big influence on how the user perceives your site. For this reason you need to test how well the key features of your site are rendered in a variety of environments. In order to create an effective testing environment you will need to duplicate a representative range of user environments. This need not be an expensive exercise; you can install a range of browsers and test on a range of monitor resolutions on a single PC.

Document your code

In addition to HTML mark-up, your site will probably make use of code, such as:

- Server-side scripts (such as CGI) that process input received from forms or applets
- Page-based scripts (such as JavaScript) that are interpreted and executed by the user's browser
- Executable pieces of code which are launched within the web environment but which go beyond the functionality built into the user's browser – for example, ActiveX components or Java applets.

All of this code will need to be documented, controlled and maintained. You should always include comment lines that describe what the code is for. In addition, if you develop the code yourself you should include a version history. You should not modify code without first creating a backup copy. Any system you have for archiving your site should include any associated code.

Before adding code to your site, ask the following questions:

- Does it work in all your target range of browsers?
- Will it be affected by the user's security settings?
- Can I replace it with a simpler solution?

Too often, on the web, the trade-off has favoured clever programming gizmos which may impress other netheads but which divert you from spending time on the real purpose of your site. Remember – you have finite time and resources; prioritise how you will use them.

Use software to automate processes

Automatically convert text content

There are various software utilities available to convert materials created in non-web environments into HTML format. If you regularly have to convert large volumes of documents held in a particular format you should systematically compare the various ways of automating the conversion. Web newsgroups and software download sites such as www.download.com can be a good starting point for this research. Be aware of the following points:

- The more consistent the original material, the easier it is to automate the conversion process
- Some conversion routines produce better results than others. For example Microsoft Word will happily filter a Word document into HTML. However, Word attempts to reproduce the existing the page layout, despite the fact that a page that has been optimised in appearance for print output rarely looks good on screen. This method of conversion results in the addition of a confusing array of formatting tags, mixed in with the content of the document, making subsequent editing more difficult than necessary.

Convert graphics in batch

If you regularly need to convert graphics from one form to another (for example converting the TIFF output from a digital camera into GIFs or JPGs) consider purchasing a batch conversion utility such JASC's Image Robot.

Validate links automatically

In addition to internal hyperlinks, many sites contain lists of links to other useful sites. A common problem faced by webmasters is ensuring that the links do not become out-of-date. Company name changes, mergers and de-mergers and changes in site structure are all common reasons why a URL might change. For this reason you need to check the validity of your links on a regular basis (say, once a month). It is important to realise that this process can be automated by purchasing one of the several site validation tools that are available. These tools systematically trawl through your site checking the links on every page and producing a list of any links that are invalid. If you need to maintain more than a handful of links, then consider automating the link maintenance process by purchasing such a tool.

Consider content management systems for large sites

If your site becomes large, consider using an automated content management tool. Most content management systems hold your raw content, such as text, images and numerical information, in a database of some kind. This information is then combined with page templates and a schematic structure in order to automatically generate the necessary web pages.

Document your procedures

You should document all the procedures connected with site maintenance. The procedures guide should cover:

- Adding new content
- Dealing with your internet service provider
- Editorial standards
- Visual style
- Graphics production
- Indexing your site
- Publicity activities.

Adding new content

Adding new content to a site is rarely as simple as uploading a new page to your web server. The new pages will need to be added to your site's indices and table of contents and be hyperlinked to other content within your site. You will need to test that the new content is rendered correctly in target browsers and has been registered with internet search engines.

Dealing with your Internet service provider

You will need to document all the issues connected with your ISP. What is the IP address and file structure of your host server? What is the password necessary to gain access to the server? What space do you have available? Make copies of all correspondence with your ISP and file them in a central place.

Editorial standards

Define editorial standards for your site as soon as you can, and educate your content suppliers in their use. Be realistic in your expectations; give your content suppliers an edited version of the guidelines and concentrate your efforts on eliminating the mistakes that cause you the most editorial re-work.

Visual style

Your site will have a recognisable visual style defined by a combination of fill colours, text sizes and fonts and graphic positioning. In order to facilitate the reproduction of this visual style you will need to document how it is achieved in a style guide. A style guide should specify the site's visual standards and how you can actually follow those standards. For example, how do you achieve the correct positioning of graphics in relation to text? You should document all of this information and collect it into a single document that you can issue to all content developers. It is important to understand that the more straightforward the instructions in the style guide, the greater your chances of growing a visually coherent site in the long term. If a content developer has to perform half-a-dozen separate operations each time they create a heading (for example, right-align, change to RGB colour 00099, set font to Arial 24pt Bold) they are less likely to apply this consistently than if they just make a single reference to a CSS file in the document's heading and then simply tag the headings H1, H2, H3 as necessary.

Graphics production

One way to create a consistent look for your page is to give all of your graphics a similar treatment and style. With photographs this will often involve passing a source file through a series of transformations in an illustration package. For example, an original photograph might be transformed from colour to black-and-white and then transformed so that it resembles a sketch rather than a photograph. A white edge-fade might be added to blend the graphic into its background. Finally the graphic would be reduced in size and optimised for the standard browser colour palette. Whatever the transformation sequence employed, it is essential that you document it in some way, so that it can always be reproduced. It is easy to assume – soon after you have created a graphic – that you'll always remember how it was done. Experience usually teaches that if you need to create some new graphics, six months after you've created the originals, you won't remember your original formula.

If an external agency creates the original graphics for your site, you should ensure that the site can be expanded without recourse to the agency. This means that, as well as providing you with a set of graphic resources, the agency should provide you with instructions on how new graphics can be created to follow the chosen style. This requirement should be part of the design brief that you present to the agency.

Indexing your site

The golden rule here would be – decide on an indexing policy before you create your first piece of content.

To effectively maintain a keyword categorisation system, it is a good idea to put in place a consistent process from the site's inception. If you are going to index your site using a keyword system, you will need to establish a policy on this before you create your first page. Although internet search engines such as Yahoo! and AltaVista will automatically index every word in your site, this does not absolve you of the need to think about your indexing policy. Free-text search can be a blunt instrument and you may want to categorise your content by keyword, perhaps for use within a site-specific search engine or an hierarchical categorisation system. You may be tempted to put off this decision until later in your site's development; after all if your site only has 20 pages of content, then keyword indexing might not seem particularly crucial. However, the thing to remember is that indexing is a time-consuming and tedious process; if you have to backtrack and categorise a large volume of legacy content, the work involved may be so daunting as to make the process unfeasible.

Publicity activities

List the publicity activities that you go through on a regular basis, for example, submitting the site to search engines, sending PR releases to the computer press, and informing newsgroups and linked web sites of any site news.

Create an archiving policy

Make a back-up of the site before you change it

Before you make significant changes to your site, you should first back up the existing site (or at the very least, the portion which has changed). There are several reasons for this:

- If there are problems with the new version you have a back-up to revert to
- The old versions of the site may (eventually) acquire a historical or cultural value. (Imagine: your successors may want to trawl back through the site archives for its 50th anniversary celebrations)
- You have a permanent record of all the material that you have published on your site – you may need this for legal reasons
- You may wish to republish a resource file, such as an old article or picture. Having a coherent archiving policy ensures that you will be able to track down the original.

A straightforward back-up method is to zip up the entire site and then copy the zip file to CD or some other backup storage medium.

Archive source materials

As well as archiving the site you should consider archiving the source materials from which the site has been assembled. This is because the process of converting original materials for web delivery usually involves some kind of filtering, in the process losing some of information contained in the original. For example, low-resolution GIF files may have down-filtered from high resolution TIFFs; HTML or Acrobat files may have been converted from Framemaker, Word or SGML sources; MIDI sound files may be edited fragments of a larger piece; WAV files might be down-sampled from a CD-quality sound source. By retaining your source materials you give yourself the option of republishing in the future at a higher resolution, or making use of some information available in the source material which is not present in the version currently published on the web. Changes in available bandwidth, browser technology, client requirements or formatting changes may mean that you will need to revisit your original source materials.

Choose a user-friendly password allocation system

There are a variety of password authentication systems available with different costs and levels of security. When considering which to choose, take into account any limitations that the system imposes on the user in terms of the password that they can choose. Password overload is now a common feature of everyday life – ideally your password system should allow users to choose something that is meaningful to them and easy to remember. Be aware that if your system automatically generates a password for the user you will receive more lost password support calls than if the user can choose his or her own password.

Don't make commitments you can't keep

How often do you find that a LATEST NEWS section of a Web site leads you to information that is months (or sometimes years) out of date? By using terms such as NEW, BREAKING, CURRENT or LATEST you are creating a commitment to update this part of their site on a regular basis. Before you do this stop to think. Do I have the resources to do this? Will there be a sufficient supply of new materials for this to work? If you are in doubt about the answers to these questions, then don't go ahead and create the section. Commitments of this kind may turn into an albatross around your neck – don't make them if you might break them.

Checklist

Priorities established		
Content supply secured	Development plan created	
	Phases established for introduction of new content	
	Agreements reached with content providers	
	Content providers trained	
	Management briefed	
Design simple	Complex text mark-up avoided	
	Style sheets used	
	Links only where necessary	
	Plug-ins and applets avoided unless essential	
Code tested		
Software used to automate processes	Text conversion	
	Graphics production	
	Link validation	
Content management options reviewed		
Procedures documented	Adding new content	
	Dealing with your Internet service provider	
	Editorial standards	
	Visual style	
	Graphics production	
	Indexing your site	
	Publicity activities	
Testing policy developed		
Archiving policy in place	Make a back-up of the site before you change it	
	Archive source materials	
User-friendly password allocation system chosen		
Commitments realistic		

11
Help system design

Matthew Ellison *is manager of training and consultancy for Digitext, the Oxfordshire-based on-line Help and Intranet development company. He is known as one of the UK's leading experts in Windows on-line Help, and divides his time between designing on-line information systems for a range of major European corporations and presenting courses on Help authoring techniques. Matthew is chairman of the annual 'European Online Help Conference'.*

Historical context

On-line Help is an increasingly important medium for presenting information on software applications to users. Over the past ten years, there has been a steady trend in PC applications away from character-based applications that ran in DOS towards GUI applications designed for the Windows platform. The strategy for documenting DOS based applications was clear: you wrote a user guide and packaged this in book form along with your product. Very few software applications had on-line Help, but there was little incentive to develop it because no-one ever used it. Why? Because it was difficult to navigate, unintuitive, and rarely answered the questions that you really wanted to ask. But this situation has changed. Many software applications are shipped without any paper-based documentation at all, and they rely solely on on-line information.

There are two reasons for this. The first, and most persuasive reason for software vendors, is the cost saving gained by not printing a manual. The second reason is the advance in Help technology that means that the experience of browsing Help is now far more rich and intuitive.

Help as an on-line manual

There is an argument that if people are happy and successful using books, perhaps we should aim to make the experience of browsing a Help file as close as possible to reading a book. However, I believe that this argument has two flaws. The first is the misconception that it is possible to simulate the experience of reading a book in an on-line medium. The second is that, even if it were possible, it wouldn't be desirable – you would end up with a medium that simply duplicates an already successful medium (albeit at reduced cost) rather than complementing and adding further value.

Compare the process of reading a book to that of browsing a Help file. When one picks up a book, although the arrangement of pages implies a reading order, one typically starts by flicking through to obtain a general concept of the size, scope, structure and content of the book. Then one might typically turn to a section of interest and start reading. However, it isn't long before, instead of reading slavishly through each page in sequence, one skips to other sections by following cross-references or simply because one's eye is caught by an interesting heading on the opposite page. When reading a book, at any one time one has a clear understanding of context and position within the book through headers and footers, as well as the physical opening of the book. Locating a specific item of information can be somewhat haphazard, although the index at the back of the book is a useful aid.

The strengths of the paper-based manual are:

- Readers are free to flick though and randomly access any page at will
- The physical nature of the manual means that readers always have a clear idea of their position within it

- Information is relatively easy to read from the page due to the high resolution and static quality of print on the page.

By contrast, users who access a Help file may be doing so context-sensitively, in which case they enter the Help directly at a point that gives them directly relevant information. If not, the vast majority of users typically go straight to the on-line index, type in the first few letters of a phrase expressing the information they are seeking, and move from there straight to that information. If their question is answered or their problem solved, they will probably exit the Help immediately and continue with the task in hand. If they need further information, they are restricted to where they go next by the links that the Help author has provided from the current topic.

The strengths of on-line information are:

- Instant availability of relevant and useful information
- An interactive index with links to appropriate topics
- Links between topics and to other on-line documents such as tutorials
- The potential for graphical content in full colour.

So paper-based and on-line media are used differently, and have clearly distinct strengths and weaknesses. By implication then, a complete set of documentation that provides reference, guidance, conceptual and instructional information should include both paper-based and on-line components. Moreover these components should be designed and written to suit the particular medium, with the aim of providing on-line Help that *complements* rather than simply duplicates the manuals.

Unfortunately however, the hard world of commercial reality often forces compromises. Authors rarely have the luxury of being able to devote time to crafting separate on-line Help and paper-based information in this way, and 'single-sourcing' is becoming one of the buzzwords of the industry. This means using a tool that enables basically the same information to be output either as printed pages or as on-line Help. Many single-sourcing tools now allow the marking of 'conditional' text that will be output to one of the media, but excluded from the other.

This avoids the appearance of phrases such as 'Click here for more information' in a printed manual.

Although the results of single-sourcing are usually far from ideal, the time and cost savings to be gained from this technique are often compelling. With careful planning, judicious chunking of information and a single-sourcing tool that allows conditional text, the resulting on-line Help system can be quite acceptable.

When do users use Help?

Perhaps due to its name, people turn to on-line Help not when they are seeking general guidance, hints or tips, but when they have a problem. Pressing F1 or clicking the Help button is often the metaphorical equivalent of shouting 'Help!!'. When users have a specific problem they usually want an answer quickly. They want to be able to go into the on-line Help, find the topic providing the solution to their problem, read and understand it, and then come out of Help and continue with the job they were doing. If they are seeking step-by-step instructions for completing a specific procedure, they may choose to remain in Help as they complete the procedure, reading each step from the Help as they complete it in the software application.

Characteristics of an effective Help file

So what does the way people use Help tell us about the required characteristics of an effective Help file? First of all, it should provide a navigation mechanism that enables users to find the topic containing the answer to their questions as quickly and easily as possible. This means that the process should be obvious and intuitive so that thinking time is minimised. In addition, it should involve the smallest number of key presses and mouse clicks.

Table of contents and index

Accessing information via a table of contents can require as little as three mouse clicks to complete. However, it usually requires the user to scan through the same number of lists of options, and to analyse and understand how the Help author has organised the topics.

Using an index, on the other hand, may require anything up to ten key presses to highlight the appropriate main index entry, followed by two mouse clicks. The difference is that the user should not have to think about how the Help is organised. They should simply be able to type the first few letters of the word or phrase that they use to express the concept, topic or problem that is concerning them. For this reason, an on-line index is usually a far more important and more frequently used navigation tool than a table of contents.

There are other reasons why the table of contents has been a little-used navigation tool by Help users in recent years. Although WinHelp 4 (the Help viewer for Windows 95 and NT 4) provides an expandable table of contents, it is available in a separate 'Help Finder' window that cannot be viewed simultaneously with topic content. Moreover, it does not track users' progress through the Help file by highlighting the current topic as links or browse buttons are used to move from topic to topic. As a result, the table of contents is not helping the user to orientate themselves within the Help and avoid getting lost and confused.

Full-text search

Full-text search is an often over-rated mechanism for accessing information within Help. Although it does have a role as a 'safety net' when users have not been able to find what they are looking for in the index, it should never be used as reason for skimping on an index, or even omitting it altogether.

The fact is that a full-text search is a comparatively blunt instrument for locating useful topics, because it will return a list of all topics that include the word or phrase being searched for, whether or not the topic provides any valuable information of that word. To give any results, full-text search relies on users searching on a word or phrase that has an exact match within the topic text. This means that users are often frustrated by getting no matches because they happened to type in a word that did not match the term used within the text.

Natural language querying

A comparatively recent navigation feature for locating specific information within a Help file is natural language querying. This means that, instead of the user typing a word or phrase into the index and hoping that it matches a keyword that the author has included, the user can simply type in a phrase or question that expresses their query. The technology will then analyse this text, and attempt to match it to content within the Help. The result is a list of topic 'hits' that contain relevant information.

This technology is good news for the Help author since a well-researched and constructed index can consume as much as 10% of a Help development budget. However, the results of natural language queries can be somewhat unpredictable and hit-and-miss with the current technology. This said, development of natural language capability is progressing rapidly and certainly promises usability benefits in the future, especially if combined with voice recognition technology. It could even mean that traditional indexes will be a thing of the past, which would have a significant impact on Help development schedules.

Context sensitive access

One of the most obvious advantages of on-line Help is its ability to provide direct assistance that is relevant to the user's 'context' within a software application. This typically operates at either dialog or field level, with field-level Help for Windows 95 and NT applications normally being displayed within pop-up windows.

In practice, context-sensitive Help is limited to describing the purpose or function of the dialog or field in focus. Because the Help has no way of knowing the precise task the user is engaged in (most application dialogs can be used for a variety of tasks such as creating, updating or deleting), it cannot provide step-by-step procedural instructions. A useful technique however is to provide links from context-sensitive topics to related procedural topics. Thus, a pop-up 'What's this?' Help topic could contain a button that, when clicked, displays a menu of topic targets that would be displayed in the main Help window.

Many Windows 95 and NT applications provide only field level 'What's This?' Help, without any dialog level Help. Ideally, both would be available, since users sometimes require Help that provides an explanation of the over-all purpose of the dialog and how the individual fields inter-relate.

Embedded Help

Embedded Help is user assistance that is built into the application interface itself, usually within a Window dedicated to Help, but sometimes even as messages or pop-up information that is totally integrated within the application interface. Microsoft has been focusing on the implementation of such user assistance with HTML Help, and a number of its consumer products (including Microsoft Works and Money 99) incorporate this technology.

There is some debate amongst Help usability experts about how far it is desirable to blur the distinction between an application's interface and its on-line Help. The advantage is that it makes it possible to place important guidance information in front of users without them having to request it. Arguments against it include the fact that sometimes users have questions that don't relate to the task they are currently performing. Such people would rely on being able to access a source of information clearly identified as being on-line Help.

Links between topics

Although users typically only need information from a single topic when they visit a Help file, links can potentially provide an important way of accessing further information. This could be:

- Increased detail on the current topic
- Related information of a different type (for example a procedure that the user can execute using the command described in the current topic)
- Information that more specifically addresses the user's current problem or query. This is particularly relevant if the initial search for information (whether through the contents, index or full-text search) has not displayed the most appropriate topic.

In this way, users can use links to 'home in' on the exact information that they need.

A useful and widely used feature of WinHelp 4 is the ability to create pop-up links to optional information such as definitions, examples, answers to questions, or even further links to related topics. Such pop-up links are not so well implemented in HTML-based Help, since there is no recognised HTML tag for defining pop-up hotspots. Moreover, it is technically very difficult with current browser technology to display HTML-based content within a window that has the same characteristics as a WinHelp pop-up.

Links, although one of the features of an effective Help system, can cause problems. One of the most common mistakes that new Help authors make is to provide an over-abundance of links between topics. When mentioning a term within a topic, there is often a temptation to provide a link to another topic that describes the term more fully, without necessarily considering whether such a link is appropriate or useful to the user. This is a particular problem when jump (rather than pop-up) hotspots are used within the body of a topic. The result can be that users are diverted from the main topic of interest. At best, this is an unwelcome temporary distraction, but it can sometimes mean that the user is unable to get back to what they were originally reading. The solution is to group such links to related topics either at the end of the topic, or within a non-scrolling region at the start of the topic.

Since the release of WinHelp 4, the most common mechanism for linking related topics has become buttons that invoke either the ALink or KLink macro. This means that when the user clicks the button, the links are resolved in real-time using keywords, and a list of target topics is displayed. This has the following advantages:

- It reduces the amount of text within the topic
- It makes it easier for Help authors to maintain the links (they can delete topics without worrying about breaking links).

However, there is no evidence to show that links of this type are any easier or more intuitive for users. Indeed, there is a risk that the user will miss out on useful links because they are not displayed as a prompt list.

Regarding the issue of selecting destination topics for links, it is often difficult to specify the ideal selection using one or more keywords. If links are particularly critical for a given Help file, it may be worth the extra time and effort required to specify each link individually and have them displayed to the user as a hard-coded list of hotpots.

Windowing

Despite the recent trend away from secondary windows (mainly due to the complexities of implementing them in HTML-based Help) they can play an important role in helping users to navigate a Help system successfully. There are three main reasons for displaying a Help topic in a secondary window, as opposed to the main window:

- To signal the role and type of the topic (for example, topics displayed in a narrow window at the right-hand side of the screen can be recognised instantly as Help for procedures)
- To enable two complementary topics to be displayed on screen simultaneously. For example, a graphic might be displayed in one window, with the text describing the graphic in another window. This would enable the text to be scrolled while still keeping the graphic in view in the separate window
- To provide additional information without replacing the current topic (enabling the user to follow a 'side-track' without losing sight of the main path).

The last of these points is important, because it can solve the problem (described in the previous section) of users getting lost as a result of clicking on jump hotspots within the body of a topic.

Secondary windows, despite their potential benefits, should not be overdone. WinHelp 4 allows the creation of up to 255 secondary windows. In practice however, most successful Help systems use no more than four or five different secondary windows in addition to the main Help window. Any more than that, and the user is in danger of being confused by the range of different window sizes and positions being displayed. WinHelp 4 also theoretically permits up to nine windows from the same Help system being displayed on screen at the same time. Most sensible Help authors would however ensure that the design of their Help systems precluded more than two windows from being on screen simultaneously.

The future of on-line Help

What are the changes in Help design and technology that we can expect to see over the next two years or so? Firstly, I expect there to be a divergence in the Help platforms on which authors will deliver their on-line user assistance. Over recent years, WinHelp has been the standard mechanism for delivering Help on a Windows platform. Now, with the trend towards HTML-based and web-based applications, authors need to be able to write on-line Help that can be read by a range of different browsers on a range of platforms. Microsoft's HTML Help is not necessarily the solution for everyone because it is not a truly web-centric Help delivery system – it is designed to provide Help for applications written for the desktop using development tools such as C++, Delphi and VB rather than Java or HTML. For this reason, there is currently a lot of interest in Java-based Help systems such as Sun Microsystems' JavaHelp or Oracle Help for Java.

It will be interesting to see how quickly Help authors embrace the technologies and techniques that HTML-based Help makes possible. Some examples of the new opportunities are dynamic content, interactive forms, richer table formatting, and links to the Internet. At present, there is a tendency when migrating from WinHelp to HTML-based Help simply to mimic the appearance and navigation techniques of the WinHelp format. Attitudes are however beginning to change with the release of mainstream products, such as Microsoft Money 99, that employ a highly embedded and radically different looking model for on-line Help. Let's not lose sight of the fact though that, although the technologies may change, it is ultimately the content of on-line Help that is of primary importance, and there is no reason to alter this as a result of migrating to a different Help platform or tool.

As a final thought, I expect to see the trend for embedding on-line Help within the application's user interface continuing and being taken to greater lengths. This will mean a need for closer co-operation between Help authors, human factors engineers and application developers, with the Help author retaining the lead role in designing and developing the content of on-line Help.

12
PDF — Pretty Darned Flash

Bryan Little *is Information Design Manager for BT (CBP) Ltd, which makes special functionality telephones and full availability telecommunication switches for financial centres around the world.*

Bryan was first introduced to Adobe Acrobat in April 1996 and has integrated it into his business practices. He uses Acrobat to proof publications and to deliver them to a global audience via the web, CD-ROM and paper.

Portable Document Format (PDF) technology has been around since 1992. It has a greater foothold in the graphic designer's camp for the pre-press environment but more recently it has started to infiltrate the corporate world.

In its purest form, PDF allows documents from different applications on one platform to be distributed to a wider audience on another platform, without re-authoring.

There is an old adage in business that says there is no such thing as a free lunch but in the electronic documentation environment this is disproved. Adobe's decision to make the Acrobat Reader freely available has been a very shrewd marketing decision, which has undoubtedly influenced its penetration in the marketplace. It shows that Adobe understands a fundamental trait of human nature — people like something for nothing. Doing away with the need to purchase initial licensing fees and to organise annual renewals has removed a major obstacle to global acceptance.

Time is money. Reducing the time it takes to do frequent chores frees you up to do the more important things. So by incorporating PDF into your business practices you can reap rich rewards and save the company money. The more common and repetitive the tasks, the more the savings.

Since PDF files are smaller than most original formats, less space is required on disks and network servers and less time is taken to transmit files over networks. Savings are made through more efficient use of communications bandwidth and lower telephone bills.

Proof reading stages can be shortened by using PDF files to exchange drafts and comments. The original text can be copied and rephrased. An individual or member of a review team can stick notes onto the relevant place in the document. The technical author can incorporate the amendments quickly into the next draft.

The production of graphic designs can also benefit from using PDF files to enable checking of artwork, especially across different platforms. The common image format retains the fidelity of the original design, regardless of whatever native application was used to produce it. Although the PDF files are small, they are still beautifully marked.

Electronic libraries of PDF publications can be created and viewed— giving exactly the same on-screen image as the one printed on paper. These publications can be distributed on CD-ROM so they can be carried around much more easily than their paper-based equivalents. Indexes can be created, enabling readers to search for occurrences of key words or phrases anywhere in the library.

You can reduce project time-scales by sending attachments over the Internet via electronic mail (e-mail). The PDF file format is compatible with web protocols and browsers, which enables you to implement amendments quickly and easily. Your community of interest can benefit from combined knowledge, across the globe.

WIIFM — what's in it for me? That is the first question you ask yourself when somebody proposes something new. People do not embrace change willingly. They prefer to stick to their nice cosy world. But how can you ignore such benefits as:

- increasing your publishing productivity
- shortening your project time-scales
- reducing your distribution costs
- reaching a wider audience, more quickly
- ... and saving the Brazilian rain forest, into the bargain.

This chapter sets out to provide a practical demonstration of the benefits to be harvested by introducing the use of PDF into your everyday working practices. PDF is not just another technology but also a business efficiency tool. If you can print it, you can PDF it.

Documentation exchange

The application of PDF to electronic mail attachments brings rich rewards. Like the equivalent of Nescafé in electronic document terms— just add water for instant results.

E-mail is essentially a text-based communications medium that is good for sending short messages around a business community very quickly. However, the formatting restrictions are limiting, since character attributes cannot be used to enhance your message. If you want any kind of formatting you have to resort to attaching a native file to the message. For many users, that means a Microsoft Word document.

Size is important and don't let anybody tell you otherwise. Size is a key element in file storage and transmission time and both of these can be related to money.

Native file type	File size (bytes)
*.txt	3
*.doc	19,456
*.ps	11,008
*.pdf	2,406

Just to send around a simple three-letter file you need to include the Microsoft Word baggage of 19 kilobytes. It is reckoned that approximately 65% of the audience only want to read the information and will discard it afterwards. What's more, most e-mail systems keep copies of the attachment for all the recipients of the message. So valuable space on your e-mail server is therefore consumed by duplicate copies of the same file, baggage and all. This means that regular housekeeping is called for, in order to minimise wastage. So why send the baggage, which you need only if you want to edit the file?

Typical savings in storage space are shown in the table below.

Document type	**Native file type**	**Native file size in Kbytes**	**PDF file size in Kbytes**	**% saving**
Product Specification in Microsoft Word (text and diagrams)	*.doc	12,130	4,338	64%
User Guide in Ventura Publisher (text and vectored diagrams)	*.vp	8,720	3,670	58%
Microsoft PowerPoint presentation	*.ppt	12,379	831	93%
User Guide in Microsoft Word (large quantity of screen captures)	*.doc	20,010	1,800	91%
Vector diagram	*.cdr	680	140	79%
Raster image	*.tif	15,176	39	99%

Sending the native file format also means that you have to have the application that created it, merely to open it to see its contents. That's OK if you are using a 'popular' application, like Microsoft Word but what happens if it is a specialist application, like a presentation, project management or database package. It is rather an expensive exercise just to view or print out another person's results. You may also need training in order to understand how to operate the application.

Up to now I have only been talking about text, but what about graphics? Can you guarantee that all your clients have Adobe Illustrator or Photoshop, CorelDRAW or PHOTO-PAINT? Whatever package you have used to create graphics, you can produce a PostScript file, convert it to PDF using Acrobat Distiller and your clients will be able to look at it.

Transmission across networks is the third key issue. The charges made by your telecommunications provider are dependent on the call duration and the distance between the two parties. A large file attached to a message will take a long time to transmit via a slow modem. Even if you have a fast modem, the transmission can be slugged by the speed of the modem at the other end. All this takes time and clocks up your telephone bill.

Adobe Acrobat uses a highly efficient compression algorithm that squeezes all the air out of a file so that you are left only with the solid essence of the message. This small file can be sent quickly and efficiently around your communications network and to the outside world. At the other end the data is inflated to show its original composition without loss or degradation of the information.

Some routers and bridges in communications networks have a restriction in the size of files that can be attached, to stop the communications highways from becoming clogged up. If your file is too large then the message and the attachment cannot pass through the gateway. The result is that your message is not delivered and you receive an error message. If native word-processor files are the cholesterol of communications networks, then PDF files are the anti-coagulant.

Document review

In the beginning there was paper and a coloured pen, usually with red ink, when it came to the task of proof reading the various drafts of a technical publication. But that was in the days before the arrival of PDF to revolutionised and streamline the process.

Picture the scene. The latest proof for review would arrive (or not) depending on the efficiency of the Royal Mail. It was not available before about 9:30am unless a premium charge was paid for a courier service. On occasions the envelope would be partially devoured by a hungry automatic sorting machine, rather than a trainee postman. Likely as not, the contents were spewing out of the envelope, soiled and disfigured.

This method needed a lot of time, patience and dedication. I would start early in the morning (and I mean early). The process sometimes took the best part of 12 hours, end to end. The medium was not reusable — black ink on single-sided white paper. The only course of action open to me was a slog of crossing out, rewriting passages and squeezing new material in wherever I could find the space. Sometimes I even resorted to a pair of scissors and a reel of sticky tape.

All that is now a thing of the past. With PDF, proof reading is done electronically. Instead of printing the proofs to a printer, a PostScript file is created and distilled into a PDF file using the Acrobat Distiller. Essentially, the process of proof reading has not changed. I still have to read the copy produced by the technical author, validate its accuracy and check that the words and pictures agree. The difference is that I am reading from the screen rather than from printed paper. These files are small enough to be written to a floppy disk. They are still subject to the vagaries of the postal service but because they are smaller than the paper equivalent they arrive in pristine condition.

So what's the difference, I hear you ask? Well if the copy is correct and the diagrams are perfect then nothing. But then, how often does that happen? It all comes down to the way you make your comments.

Spencer Sylver — now there's a name to conjure with! I am sure you are all familiar with the British chemist who accidentally produced the non-sticky adhesive that made the Post-it® note one of the great discoveries of the 20th century. Adobe Acrobat has implemented a similar concept called *notes* for PDF files. Notes allow you to make comments as you review a document and to attach them electronically without altering the original.

'Stickies' (as I call them) have an immediacy all of their own. You don't have to go looking for a scrap of paper or your workbook, or open up an electronic scratchpad. You can do it there and then. The variable size of notes means that they obey Parkinson's first law. They can expand to fill the space required, up to a maximum of 5000 characters. I contend that is more than adequate space for the comments of a verbose reviewer.

Because the information in the original document is held electronically, it can be reused by simply using the text selection tool to copy and paste the text into the note. You can juggle around with it to your heart's content until you get it exactly right. Better still, the technical author can copy your words verbatim, by cutting and pasting the amendments back into the native application. Both the reviewer and technical author save the time of re-entry and eliminate the error-prone process of copying. If you have a repetitive construct that needs amendment, you can swipe the text, select all, cut and paste, and then change the variable element.

Once the note has been completed it can be deflated so that it appears as a small coloured page icon. You can locate the note anywhere on the page so that it does not obscure the appearance of the original document. You can put it on a particular word or in either margin. Beware, when you place it outside the page boundary it can be missed (believe me). You can re-inflate the note at any time to read its contents. No matter how bad your handwriting, you can guarantee that your comments can be read.

Committing your comments to notes provides an added bonus in subsequent drafts. All you have to do is display both PDF drafts in the same Acrobat Exchange window, using the Tile Vertically option. Place the previous draft on the left-hand side and the new draft on the right-hand side. Notes can be found quickly, without having to search through every page. A keyboard shortcut (CTRL+T) or menu option will whisk you to the next note in sequence. Bring up the corresponding page in the current draft and it becomes a simple process of *spot the difference*. You can quickly check that the changes have been incorporated correctly. Simply repeat the process until all the notes have been validated.

There is some merit in developing your own dialect of shorthand for comments in 'stickies'. For example, you can use the square bracket characters '[...]' to encapsulate your remarks to the technical author to minimise the ambiguity between text destined for document content and your publication instructions. You can easily collate the notes you have added, because they are not part of the original document. All the notes can be summarised into a separate PDF file, with each note containing the page, number, date, time and its contents.

In a group review situation, the ability to change the preferences for notes, in respect to label, colour, font, and point size provides a method of identifying who made the comment. Rather than summarising the notes, you can choose to export the notes to a separate file. Once you have received the files from all your reviewers, you can import them into a single document and see all the comments in the same environment. This saves time rationalising duplicate comments.

Using PDF files to proof read publication drafts can therefore save time for you and for the technical author. It can speed up the proof reading process and reduce the project time-scales. My perception is that it now takes me half the time to proof read three drafts of a publication. Best of all, these PDF 'stickies' are free! No matter how many you use, they have no effect on your stationery bill.

Cross platform proofing

Because PDF output looks identical to the printed form, the benefits are not just limited to text publications. It can be used to proof read intermediate drafts for graphic projects as well. If your publication requires some element of design then you are likely to call upon the services of a graphic designer. The graphic design fellowship has gravitated towards the Apple Mackintosh (MAC) platform while the corporate community has chosen the Microsoft Windows PC platform...and never the twain shall meet.

In the past you had to rely on large amounts of soft packaging, non-standard sized parcels and the Parcel Force. Many of the production processes used to create a colour proof were time consuming, laborious and expensive. Disproportionate amounts of padding were needed to protect this delicate medium from the ravages of transport and the attendant manhandling. The result looked like a cotton-woolled quarterback guarding against a sacking from the opponent's offensive line. An extra inconvenience was that the parcel had to be taken to a local Post Office or collected by over-night courier.

The problem is one of interfacing between the different platforms and this is where PDF can help. PDF is platform independent – a kind of Esperanto of computing systems. The trouble is that you are both comfortable in your own environment and unwilling to consider the problems of the other's platform. Each of you has your own set of assumptions, different vocabulary and jargon, unique creative applications and personal expectations. This is a potential communication problem, just waiting to happen. The other thing about designers is they tend to live in the most out of the way places and usually work at home. The distance between my place of work and the home of my designer is about 150 miles, with the M25 as an integral part of the journey. On a good day the journey can take up to three hours.

The raw copy I create on my corporate PC is generally in Microsoft Word and my rough diagrams are produced in CorelDRAW and Corel PHOTO-PAINT. Originally, I transferred a hard copy using the postal service or via facsimile transmission. The graphic designer's favourite application, Quark XPress, was used to create the graphical image, so all the information had to be keyed in again on the MAC – a lengthy and error prone process.

If the graphic designer wanted to send me proofs they had to resort to using the same methods in reverse. A facsimile machine is not a very good medium for checking graphical proofs. The print is black and white, invariably too dark, lacks clarity and does not provide enough definition. Colour is out of the question. This procedure incurred time delays, required expensive courier service and was constrained by collection deadlines.

Adobe has been in the MAC environment for a number of years, providing graphic designers with such tools of their trade as Photoshop and Illustrator. So PDF was not a complete unknown. But in the Windows PC world it is still a relative newcomer. Initially, a floppy disk was the medium to transfer PDF files between the two platforms. We were using different versions of the Adobe Acrobat suite to create and view the PDF files. The designer used Acrobat Distiller to convert the PostScript file output from Quark XPress and Acrobat Exchange for post publishing tweaks such as cropping the pages to size, creating the thumbnails and annotating the artwork.

They say that '*a picture is worth a 1000 words*' but what format do you choose – raster or vector images. Raster images give you a fixed density, which compresses to a smaller file size. Each pixel is uniquely defined so that it displays and redraws quickly. If the pictures are only going to be viewed on screen, then it may be acceptable to decrease the resolution of the raster graphics. However, the image may appear blurred when you zoom in, becoming pixelated at high levels of magnification. On the other hand, vector diagrams are scalable and retain the clarity on zooming. The detail is contained in print descriptive language commands, which means a larger file size and a longer time to redraw. You can minimise this effect by avoiding such things as embedded raster images, fountain fills, textures and blends. Every time you move around the image the diagram has to be redrawn. You can get round this problem by exporting the vector diagrams as a raster image but you sacrifice the clarity of the image when zooming. Beware of clip art that has too many objects. The drawing may look very professional but take an age to render.

I review the contents of the PDF file on my own first. Here I am able to increase the viewing ratio to the level of magnification that is comfortable to examine the detail of the design. If necessary, I can use the Hand tool to move around the design without changing the magnification ratio. I use Acrobat Exchange to add 'stickies' with my comments or queries at the appropriate place on the design. These act as prompts for asking questions during the subsequent discussion with the designer.

The bonus I find is that I can have a conference call with the designer in real time to discuss the progress of the design. Although we are miles apart and operating on different platforms, we are both able to use our own variant of the Acrobat Reader or Exchange to look at the same image. If the changes are only minor, they can be incorporated in a matter of hours and reviewed again. This reduces the pressure on your project management skills as you juggle with rapidly decreasing time-scales the closer you get to the completion date.

However, despite all the benefits of electronic communication, I find it is still essential for a face-to-face meeting with a designer at the start of a project, to establish the concept of the design. These meetings get you out of the office, you are able to eat a good lunch and you clock up the requisite number of business miles.

Electronic library

Back in the days of large paper manuals you could always spot one of our technicians by their Neanderthal posture and the way their knuckles dragged along the ground. This condition had evolved over a period of years and was caused by carrying around a load of bulky manuals under one arm. The original user guides and technical manuals started off during the early days of desktop publishing in the classic A4 size, they weighed a ton and were difficult to fit into a briefcase. If this trend had continued, you would now have to tote around a jumbo-sized suitcase. Even though these publications were subsequently reduced to a natty A5 size, they still presented a logistical problem. Something had to be done and quickly. Internally, I had maintained a selection of titles in a publications library but these would often be 'borrowed' for some urgent need and never returned.

I had flirted with the idea of distributing publications electronically before, but I had always run up against the problem of selecting a suitable proprietary viewer and the issue of a end user licensing. At a Corel User Association meeting, somebody demonstrated that Corel Ventura could export something called PDF marks. This meant absolutely nothing to me at the time. Using the print engine, they created a PostScript file, which was fed into the Acrobat Distiller. Out of the other end came an electronic publication in PDF, which could be viewed using the free Acrobat Reader by the target audience wherever they were, anywhere in the world, on whatever platform.

A Help desk scenario is one in which the application of PDF is a positive boon, since the image is the same, regardless of the delivery medium. The computer operator is able to look at an image on the monitor screen, which is the same as that seen by the technician looking at an image on a piece of paper, somewhere in the desert of Dubai. They can have a meaningful dialogue because they are both looking at the same image. Translation into a serial communication of words, with all its attendant interpretation pitfalls, is circumvented. If the remote technician does not have access to the publications, the relevant pages can be sent using a facsimile machine.

The PDF specification has an inherent compression algorithm, which is highly efficient when it comes to packing the data into the least number of bytes. Even proprietary compression applications find it difficult to make additional space savings. This means that single pages or even whole publications can be attached to an inter-company e-mail message. Their small size enables them to squeeze through any network gateways.

I found that it was all very well providing these publications on an ad hoc basis. However, some semblance of order was being called for whereby a central repository of electronic publications was available to everybody in the company. This meant they could have access to product information at any time of the day, without the recourse to expending budget resource or the intervention of another person. Delays attributable to printing and distribution could be removed. Storage space on a network server was allocated for the creation of ELib – the electronic library. A PDF catalogue file was created to divide them into categories and provide hyperlinks to the electronic publications themselves.

It is important that all electronic publications have a similar appearance to the readers. If different technical authors create the company's publications, it is very easy for minor subtle variations to creep in. This highlights the need for a set of standards for electronic PDF publications. Such things as documentation information, open options, security attributes and loading indexes.

The addition of bookmarks adds value to your document and makes it easy for your readers to navigate around it quickly and easily. Bookmarks are equivalent to the contents list in a conventional paper-based book. You can choose to open a document with the bookmarks and document visible at the same time, in adjacent frames.

However much you try to assure people that it is alright to read off a monitor screen, they may still want to print out the whole document. This is because during their upbringing they have become dependent on paper. It's a bit like Linus' dependency on his blanket. If they are going to print the publication, then you had better make it look professional. Add some form of skeleton corporate covers to encapsulate the book block, and include a copyright statement. If you want conventional covers, you can cut the cost of printing and distribution by enabling it to be printed remotely. Just include the artwork to permit the printer at the distant end to create the colour separations.

What happens when you have a key phrase or expression you wish to locate but you do not know where to look for it? The answer is to index your documents using Acrobat Catalog. This allows you to create an index for a collection of PDF documents. By using the search command, the reader can then carry out a full-text search over a range of publications in the library. It is advantageous to create lots of small indexes, rather than one huge index. Limiting the indices to be searched can reduce the time it takes to locate a keyword.

The location of the electronic library on the local network means that access is limited to those people connected to the local area network (LAN) in the office. What about the agents, distributors and technicians who are out and about in the field? This is where the CD-ROM comes into its own. These days, most portable PCs are equipped with a CD drive as standard. Alternatively, you can copy an image of the electronic library onto the hard disk drive, providing you have enough free storage space. The Acrobat Reader can be copied to a separate directory so that the CD-ROM comes complete with the wherewithal of viewing the publications. CD-ROMs are relatively cheap to produce and can be re-issued on a regular basis. They are small enough to slip into a coat pocket yet can hold a complete suite of product documentation.

A final bonus - with Adobe Acrobat, *colour is for free*! Even if your company may consider it uneconomic to print their documents in colour, your readers can benefit from seeing your pictures and diagrams in glorious Technicolor. So get into the habit of creating them using coloured lines and shading.

Web distribution

PDF and the Internet combine very well. PDF is the second most common format on the web after HTML. Many companies place documents on their web site that are destined to be printed – documents such as brochures, catalogues, reports, user guides and technical manuals. The relationship between Adobe and the browser companies is so close that the browser can be configured so that Acrobat Reader fires up inside the browser window. This means you can view the PDF document without having to open a separate window. The applications I have covered in the previous sections are greatly enhanced with the addition of the Internet interface. This section describes some additional functionality that has been realised.

All the same savings that apply to inter-company e-mail systems, both nationally and internationally, also apply to web distribution – less storage space, shorter data transmission times and lower network traffic. What is more, the advent of the Internet has widened the coverage of business and personal connectivity to anybody in the world. PDF files can be attached to e-mails without the need to *zip* them up first. They can be detached quickly once they arrive and accessed immediately, without any intermediate processing or translation.

The introduction of the Internet to both proof reading of text and graphics has shortened time-scales for projects. It means that drafts and comments can be interchanged in a shorter period of time. There is no need for any other parties to be involved with the exchange. All you need to do is to compose an e-mail message and attach the PDF files.

ELib – the electronic library on a local network – has evolved into VELib – the virtual electronic library on the Intranet. The local area library has become the international library, behind the protection of the company firewall. Anybody in the company, anywhere in the world, has instant access to the latest publications. Once the new titles are stocked on the shelves of VELib and the catalogue has been updated, they are available for all to share. The immediate benefit is the immediacy. You can guarantee that your readership has access to the very latest information. Small amendments can be implemented straight away – a sort of instant, on-line publishing. Suppose somebody reports a misprint or there is a change to a data specification. No need to wait until the next update. Put it into effect there and then. The Text tool can be used to correct typographical errors on the fly, providing they do not spill over a line. Additional information can be compiled on separate pages and inserted or replaced using Acrobat Exchange. 'Stickies' can be attached to pages. This means that everybody can benefit from an individual's experience, anywhere in the world. For example, the acquired knowledge of technicians in Japan during their morning can be shared with their colleagues in Europe and the United States later the same day.

Optimisation

In earlier versions of Adobe Acrobat, you had to wait for the whole file to be downloaded before the first page was displayed. This limitation was rectified with the release of version 3. Font information is now interspersed with the content, enabling progressive rendering of text fonts and images, so that the information can be delivered a page at a time over the web. When you request a page, it appears quickly because only sufficient data to be able to display that page has to be downloaded. Whilst you are reading the contents of the page on the screen, subsequent pages are downloaded in the background. When you jump to the next page, all the data to display it will most probably be available. PDF files that are destined for the web should be saved using the optimisation option. It is located in the Save dialog box and should be the last action you do before you post your documents on the web.

Acrobat Exchange provides you with two levels of protection by the use of passwords; the open password and the security password. The open password enables you to limit access to the PDF file. Unless the reader knows the password they cannot open the file. Individual files or groups of files can have different passwords. The security password allows you to protect the PDF file in the following ways:

- **Printing**
  Disables anybody from making a paper copy

- **Changing the document**
  Prevents unauthorised alteration of the contents
- **Selecting text and graphics**
  Rules out cutting and pasting words or pictures
- **Adding or changing notes**
  Precludes the attached notes being modified.

Once you have entered the security password you are asked to confirm it. Remember to record the security password somewhere safe so that you can recall it quickly in six months time.

The full-text search functionality on the web is devolved to independent search engines, which are capable of performing a comprehensive search on PDF files as well as just HTML documents. The results of the search criteria are displayed in the browser window.

You can reduce the size of the PDF file if you choose not to embed the corporate fonts or use sub-sets of fonts where only a limited number of characters are required. In this case you will have to rely on the system fonts installed at the distant end. Additionally, you can use the Runfile option to distil and combine multiple chapters, using a single set of common fonts.

Refrain from publishing a PDF file with the creation of thumbnails, unless there is a real need for it. They are more appropriate for publications that are destined for distribution by CD-ROM. You can save space in the order of 3.5K per page for each thumbnail. Use bookmarks instead, they have a greater significance.

Publishing PDF files on the web is a matter of balance. It is a trade off between the size of the file and the quality of the image. Which is more important – the time it takes to download a file or the appearance of the graphics and the crispness of the text? It is a value judgement that will depend upon your audience's operating environment.

Conclusion

Portable Document Format (PDF) has established itself as the universal document exchange format. This chapter has demonstrated a range of uses for PDF – an attachment to an electronic mail messaging system, an integral link in the document review chain, the visual inspection of graphical images across computer platforms, the distribution of electronic publications on CD-ROM and the wide accessibility of corporate information on an enterprise-wide intranet.

PDF enables you to continue to create documents using your favourite authoring applications and to distribute them many ways. The Adobe Acrobat suite makes electronic publishing a reality. One file format means you can publish across platforms using printers, electronic mail systems, corporate communications networks, CD-ROM, and the World Wide Web.

From the simplest memos to the richest colour brochures, Adobe Acrobat lets you publish and manage your documents electronically. It preserves the original look and feel of the publication, independent of computer platform or distribution medium. This can be achieved cost effectively with the minimum amount of effort.

Incorporating PDF into your business practices makes you rethink how you work and can provide savings in both time and money throughout the whole of your business operations. You can streamline your production process, improve the service to your customers, reach a wider audience and gain the competitive edge. PDF improves the way you do business and reduces your costs. Your imagination is the only limitation.

Acknowledgements

The author wishes to thank colleagues, friends and relations in the production of this chapter.

13

Managing documentation projects

Paul Warren, FISTC, *former vice-president of the ISTC, has been involved with Technical Communication throughout his 32 year career. He has worked for Ferranti and GEC in senior publications management roles. His activities have also included Patent drafting and Patent Law, electronic publishing, and document management system specification and implementation, including work process analysis. Currently Paul is employed by Bechtel Limited as Services Contract Manager and is currently reviewing document origination, handling and reproduction strategy.*

Introduction

In the coming pages you will find some general pointers and thoughts on project management. These thoughts are not specially related to documentation production, but can be applied to project management generally. Project management is particularly about the successful and efficient management of resources within a particular business environment, and this chapter concentrates on this aspect. You are left to interpret the suggestions according to your own situation and the type of project you are dealing with. The project may be small, large, covering a writing assignment or encompassing the whole gamut of skills available to the documentation professional, from document planning and design, to the delivery of a series of printed and bound volumes. I have made very little mention of points, styles, illustrations, editors, artwork, ISDN, tif, pdf, ISO, JSP and so on. What this chapter is *not* is a technical guide; that sort of guidance should be available throughout the rest of this Handbook.

Due to space constraints you will find 'buzz words', rather than extended explanations, which you can use to hunt down more detailed information on the areas covered. I have deliberately kept the content simple and oriented it towards the tools and skills that, in my experience, I feel will be particularly appropriate and useful.

My plan is to take you through a typical scenario for a new project and point out some of the features, tools and skills that are required to set the project up successfully. With the tools and skills in place, the documentation professional should, through regular review and adjustment, be in a position to bring the project to a successful conclusion.

Getting started

Invariably we will be asked over the telephone, 'Can you call by to discuss a new marketing leaflet I want?', or be summoned by an Internal Memorandum, accompanied by a voluminous Request for Proposal, which demands, 'Attend a kick off meeting for the new Wonderbird project' and 'Please review the RFP before attending the meeting'.

Full and proper preparation for the ensuing encounter will involve gathering as much information as possible about the prospect and making some assumptions and judgements based on that knowledge.

Hopefully we will know the client, for example, a direct client for the individual practitioner, or an internal or external client for the corporate documentation manager. We will have reviewed any work already proposed or carried out for the client, and ascertained the reasons an order was (or was not) placed, together with a feel for how pleased the client was with completed projects. As part of an ongoing business relationship we will have requested some feedback on these aspects. With this information to hand there should not be too many surprises at the kick-off meeting.

It is important to ascertain just as much technical and financial information as possible about a potential new client. The individual practitioner needs to do this leg work himself through business or credit agencies if resources allow and of course via any 'trade' contacts – or rely on intuition if the risk warrants it. The corporate documentation manager will probably not have to worry too much about this aspect, as it is more often a function undertaken by the corporate business development team.

This acquired intelligence will help in making the decision on whether to do business or not – not doing business is an option for the sole practitioner, but is a decision which has normally been taken well in advance in the corporate scenario.

To summarise, enduring business relationships rely on knowing your client and satisfying their needs.

Business relationships and project planning

Having taken the decision to proceed, next follows the tricky task of determining precisely what the client requires, when it is required, and how much it will cost.

For the individual practitioner the first visit to the client, with the idea of getting a clear brief, invariably ends up as a discussion of options, ideas, and processes and the dispelling of any technical or other misconceptions the client may have. This meeting will provide the practitioner with the necessary information to enable a specification of the finished product (a *Scope of Work*) to be written, against which to calculate a price. This specification will also enable an assessment to be made of the dates for delivering the job and whether it can be finished within the timescales set by the client.

The calculation of cost will be derived from an amalgamation of the practitioner's own labour plus services bought from other resources. Sometimes the cost will simply be the cost of services accumulated to complete the job plus a mark up to cover the practitioner's management and administrative input to the project. In any event it is important to have a clear idea of who is going to do what, and when the various 'tasks' are to be undertaken. The deliverables and timescales will need to be specified and agreed with subcontractors. The details of what is going to be delivered to whom and by when, and the constraints there are on making these deliveries, will need to be documented. Others items to agree include the price to be charged by each subcontractor to fulfil their tasks and the way in which any unspecified potential (but reasonable) extras will be handled and costed. This all needs to be clearly documented in formal correspondence and summarised for ease of digestion and presentation. A knowledge of contract law and management can be helpful as projects become more complex and at some stage resort to formal contracts will be required.

Each individual task to be undertaken can be summarised by *events*, *activities* and *milestones*, and expressed diagrammatically, to show how the individual tasks link and integrate together. A particularly useful software tool for this is Microsoft Project ™.

Formal agreements (or contracts) with subcontractors and the delivery requirements and promises also form, in part, the basis for evaluating their progress. These agreements can become complex and often include penalty clauses that, for example, reduce payment for late delivery.

Having calculated the price to the client, based on the total of subcontractor costs and the practitioner's own time and costs, the information must be presented formally to the client. The price quotation or cost estimate needs to state clearly the technical specification of the delivered product, the major milestones (significant events) that are the responsibility of the client, how much time is allowed for completion of the associated activities, and any penalties and delays incurred in not meeting the milestones, together with new anticipated delivery dates if appropriate. Subsequently, if the price and other conditions are right, the practitioner will be awarded a *contract* to undertake the work. The level of formality of the contract can vary enormously, from a handshake to a multi-page legal tome, depending on the sophistication and requirements of the practitioner and client.

As the practitioner develops skills and an understanding of clients and suppliers, the process of constructing a quotation or bid will become easier. Reasonably accurate prices and timescales can be provided from standardised costings – at least to provide rough order-of-merit costs, which clients invariably want, to help them make up their minds about whether to proceed with a project.

The corporate documentation manager will be going through much the same sorts of processes except that, in this case, either totally in-house or a mixture of in-house and subcontract resource will be used. There will almost certainly be one or more technical specifications defining just about every factor possible relating to the final deliverable (from typeface to binding and the electronic format for the files prepared) including the range of documents to be produced. The cost estimate will typically be calculated from in-house man hours, external man hours and costs, plus equipment and other costs incurred specifically to undertake the specified tasks – for example, a particular version of publishing, word processing or drawing package may be required to prepare the deliverable.

Typically the documentation manager will estimate the size of each particular document, considering the content in terms of illustrations, authorship, editing requirements, revisions, printing and binding, and prepare a resource requirement in terms of man-hours and costs. The accumulation of each of these individual calculations forms a cost estimate.

The technical specification will probably define when documents are required, and the availability of labour also needs to be considered. It cannot be assumed that all the labour needed will be there just when it is wanted. Each of the individual documents is *scheduled* (planned in detail, just like a mini project plan) and other critical activities factored in to give a true picture. For instance, the availability of some specially manufactured paper, specifically requested by the client, may be a key issue.

For in-house resources a resource requirement can be established (how many people are needed and when), by accumulating the requirements for individual documents into an overall requirement. This resource requirement will inevitably have peaks and troughs, but the objective is to maintain a steady or smoothly changing workforce to achieve individual document delivery dates as well as overall delivery dates. The objective is to have the available workforce fully active and to be able to plan resource requirements effectively.

Peaks and troughs are eliminated by advancing (adding more resource to) or retarding (taking resources from) particular tasks.

Once it is established what is going to happen and when, a *Project Cost Plan* can be developed which will indicate the planned usage of internal and external resources and other costs. For a project with a long duration this could be the tool to determine the payment schedule, to give contractors a staged income in return for their efforts. Corporate organisations will have established rates for calculating the costs of the various resources used.

So in the corporate scenario we develop a *Project Plan* (what is going to happen when), a *Resource Plan* (the human resources we are going to use and when, to achieve the Project Plan) and a *Cost Plan* (the planned cost expenditure including the costs of all internal resources, sub contract costs and other costs, and when they will be incurred).

In complex situations more than one level of plan may be developed. For instance, organisations where the documentation project is part of, say, an overall engineering project, the detailed Documentation Project Plan may be considered as a *Level 3 Plan*, the start and end dates and major milestones of which are rolled up into an *Engineering Level 2 Plan*, which is shown in an overall *Level 1 Project Plan*.

An example project plan

The table and illustrations below show a basic Project Plan for a very simple 20-week task – two manuals that require the use of authors, illustrators and editors. It is assumed that the project is completed at a single site and that the staff are interchangeable between the tasks relating to the two manuals. The two tasks are required to be completed together.

First, the Project Plan is developed to determine when tasks are required to commence and be completed.

A first pass is made to determine the duration of the project and the resource requirements for each activity.

The total resource requirement can then be established.

As the project is only constrained by the completion date, the manpower can be smoothed to provide a smooth take up and reduction of staffing which provides the required man weeks of effort to complete the tasks.

By using the smoothed resource requirement and known charge rates a Cost Plan can be developed, which can be used to illustrate the weekly expenditure required and also to provide an illustration of the cumulative anticipated expenditure. Other costs (software, equipment, consumables and so on) can be added in to provide a more comprehensive cost profile for the project.

Basic Project Plan

Weeks into project	1	2	3	4	5	6	7	8	9	10	11	12	13	14	15	16	17	18	19	20
Operating Manual (Manpower)																				
Author	3.0	4.0	4.0	4.0	5.0	5.0	4.0	3.0	3.0	2.0	1.0	1.0	1.0	1.0	1.0	1.0				
Illustrator			2.0	2.0	2.0	2.0	1.0	1.0	0.5	0.5	0.5	0.5	0.5							
Editor		0.5	0.5	0.5	1.0	1.0	1.0	1.0	0.5	0.5	0.5	0.5	0.5	0.5						
Parts Manual (Manpower)																				
Author					0.5	1.0	1.5	2.0	3.0	3.0	3.0	3.0	3.0	3.0	2.0	1.0	0.5	0.5	0.5	0.5
Illustrator					0.5	1.0	1.0	1.0	1.5	1.5	2.0	4.0	2.0	1.0	0.5	0.5	0.5	0.5	0.5	0.5
Editor							1.0	1.0	1.0	1.0	1.5	1.5	1.5	1.0	1.0	1.0	1.0	0.5	0.5	
Total manpower requirements																				
Author	3.0	4.0	4.0	4.0	5.5	6.0	5.5	5.0	6.0	5.0	4.0	4.0	4.0	4.0	3.0	2.0	0.5	0.5	0.5	0.5
Illustrator			2.0	2.0	2.5	3.0	2.0	2.0	2.0	2.0	2.5	4.5	2.5	1.0	0.5	0.5	0.5	0.5	0.5	0.5
Editor		0.5	0.5	0.5	1.0	1.0	2.0	2.0	1.5	1.5	2.0	2.0	2.0	1.5	1.0	1.0	1.0	0.5	0.5	
Smoothed manpower requirement																				
Author	3.0	4.0	4.0	5.0	5.0	5.0	5.0	5.0	5.0	5.0	4.0	4.0	4.0	4.0	3.0	2.0	1.0	1.0	1.0	1.0
Illustrator			1.0	1.0	1.0	2.0	3.0	3.0	3.0	3.0	2.0	2.0	2.0	1.0	1.0	1.0	1.0	1.0	1.0	1.0
Editor			1.0	1.0	1.0	1.0	1.0	1.0	2.0	2.0	2.0	2.0	1.0	1.0	1.0	1.0	1.0	1.0	1.0	1.0
Rates (£/week)																				
Author	480																			
Illustrator	400																			
Editor	450																			
Costs per week (£k)																				
Author	1.44	1.92	1.92	2.40	2.40	2.40	2.40	2.40	2.40	2.40	1.92	1.92	1.92	1.92	1.44	0.96	0.48	0.48	0.48	0.48
Illustrator			0.40	0.40	0.40	0.80	1.20	1.20	1.20	1.20	0.80	0.80	0.80	0.40	0.40	0.40	0.40	0.40	0.40	0.40
Editor			0.45	0.45	0.45	0.45	0.45	0.45	0.90	0.90	0.90	0.90	0.45	0.45	0.45	0.45	0.45	0.45	0.45	0.45
Total cost per week (£k)																				
Total	1.44	1.92	2.77	3.25	3.25	3.65	4.05	4.05	4.50	4.50	3.62	3.62	3.17	2.77	2.29	1.81	1.33	1.33	1.33	1.33
Cumulative cost per week (£k)																				
Total	1.44	3.36	6.13	9.38	12.63	16.28	20.33	24.38	28.88	33.38	37.00	40.62	43.79	46.56	48.85	50.66	51.99	53.32	54.65	55.98
Total cost of project team (£k)	**55.98**																			

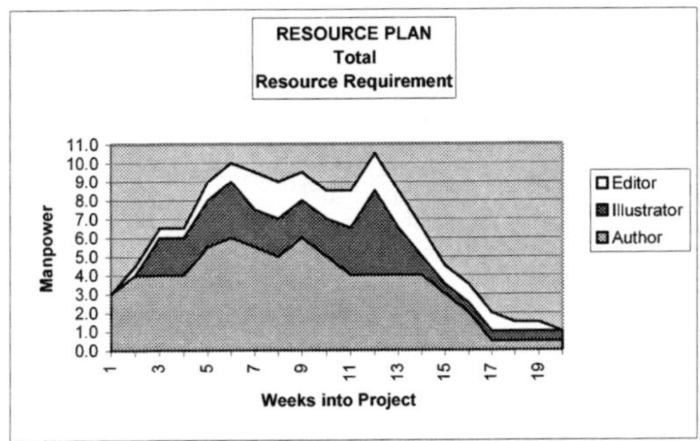

Resource Plan - total resource requirement

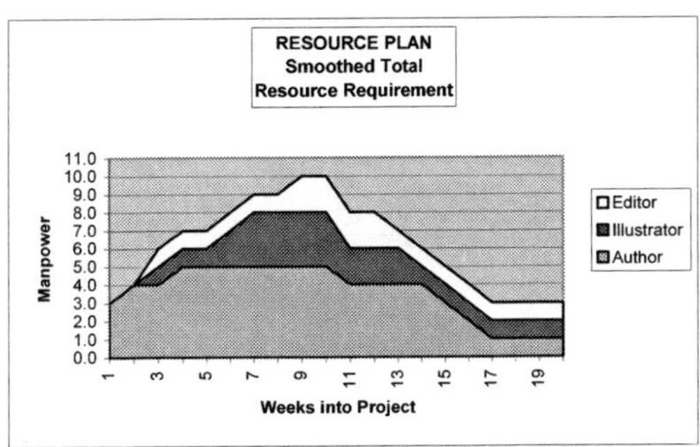

Resource Plan - smoothed resource requirement

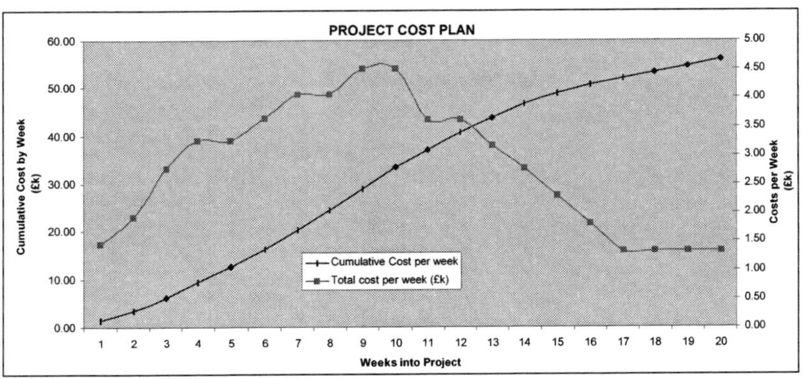

Project cost plan

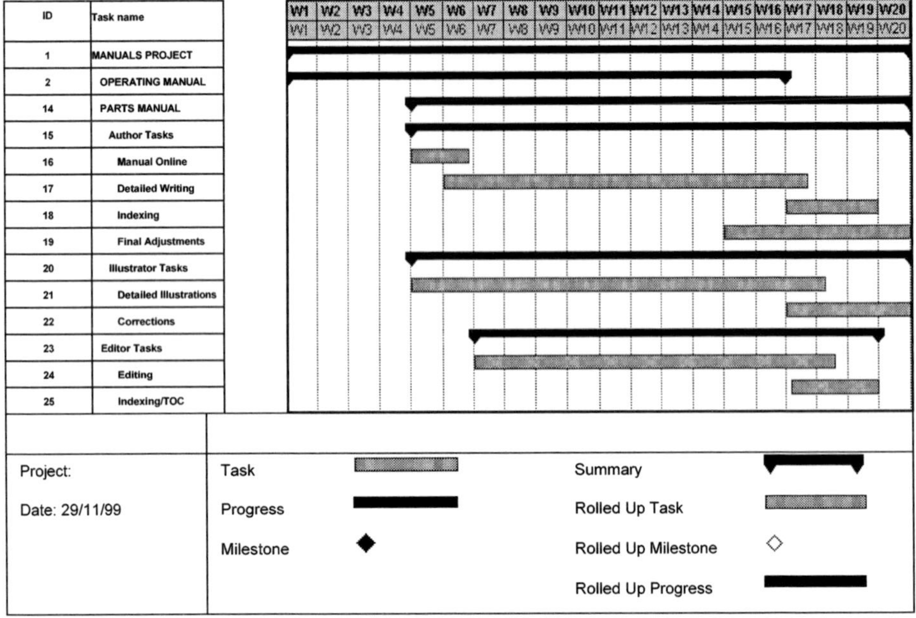

Project plan

The production of an illustrative project plan can be undertaken during the resource planning exercise. In reality the Project Plan and the resource requirement will be a compromise. For illustration purposes the Project Plan shows the original plan for production, before resource smoothing. Again, in reality, the plan will be adjusted to allow for the smoothed resource and the resource requirement will be tested to ensure that the individual tasks can be completed when required. A further adjustment to the smoothing may be necessary.

The plan for the Parts Manual is shown expanded to indicate some level of detail, whilst the plan for the Operating Manual is shown rolled up.

Scheduling and project planning tools

Today, automated scheduling and project planning tools are available which enable resources and timescales to be automatically allocated and which also have the ability to be able to smooth and reschedule labour requirements. One such tool is Microsoft Project™ and knowledge of this package would aid the understanding of this subject.

For large projects a client will probably wish to see how labour is being utilised and an organisation chart may additionally be required.

An individual practitioner (to their direct client) or corporate communications professional (to their Project Manager and later direct to the overall client) will undoubtedly be expected to fully justify how they arrived at the costs and resource requirements and understand all of the factors influencing their decisions. Invariably the client does not like some aspect of the price or the schedule or some detail of the quotation or planned execution strategy and a rework of the quotation and plans may be necessary. Document all the changes; retain the old plans and know the factors that have contributed to the changes. It is not unknown for a client to ask for justification at the end of a project.

Project manager's focus

The plans produced will clearly illustrate to the Project Manager what the project has to achieve and when. The Project Manager must have, as the primary objective, the ultimate goal and along the way the intermediate goals established in the plans. Nothing must be allowed to deviate the Project Manager from driving the team forward to the final goal and those intermediate goals along the way, re-focusing and motivating the team as circumstances change and setting new goals and targets to move them on.

At the same time, the Project Manager really needs to be realistic and not to react unnecessarily to every small setback and change which arises. It is necessary to have a broad understanding and vision of the business and technical processes that are part of and surround the project, so that questioning and coaching on future activities and direction can be carried out.

The Project Manager is the leader of the team and therefore needs to be able to demonstrate flexibility in approach and the ability to ride out or react to external and internal influences on the project, demonstrating the desired approach to the team

What do changes mean? – money!

Let's now assume that we have successfully bid and negotiated for a job and that it is underway.

We all have clients who change their minds and, once a job has commenced, particular care needs to be taken with changes in scope. Every change needs to be documented and the effect of the change both on the cost and the project schedule needs to be agreed with the client.

Changes of scope mean changes of plan, mean changes of resources and cost. These changes may be just whims on the part of the client – 'I think I'd like the red background better' or 'Can you run me another set of proofs?' – or changes due to external factors such as, for example, the price of oil going down (now there is less production, so we need to describe how we shut down or start up only half of the plant in a new section of the refinery operations manual).

Change requests should consider and show the effect on delivery, timescales, resources and cost. I am sure by now you are thinking of examples just as quickly as I am. (Another proof adds four working days, which means the printing can't be finished before the print works shut down for the summer break. Another section is another chapter and the tabs for the binders have already been delivered; some will need reprinting).

The effect of changes needs to be presented to clients in the same way as the initial quotation – clearly documented and with a request to approve before proceeding. I know life is not that simple but that ought to be the rule that we all abide by although, of course, in all business situations we must apply judgement. One thing is for sure: none of our organisations are charities and we should not give away our profit lightly.

Unapproved changes affect profit – they cost us money.

Quality

Each task that we encounter on a project will have quality criteria associated with it – that is to say, the output of the task must meet certain criteria before it is accepted for input to a further task or final delivery to the client. It is essential that, before the start of the project, the Project Manager has in place the quality criteria (physical characteristics, timeliness and quantity) and checking processes that are appropriate to each major task and sub task. Quality is measured during an *inspection* task, which may be formal or cursory. The inspection should be recorded to provide historic evidence of the satisfactory completion of the task. Verbal assurance should not be accepted as evidence of completion.

Adherence to quality standards is an essential if organisations wish to become ISO approved. These quality standards need to be fully documented and implemented throughout the organisation.

Project procedures

Particularly in the corporate arena the plethora of instructions, quality criteria, processes, timescales and so on that are required to be adhered to as a project gets more complex are best managed through project procedures. Each procedure should include a definition of the scope of the procedure, who the procedure is intended for, what assumptions are made, any cross-referenced documents and instructions to the user. These procedures can themselves form the criteria against which quality or adherence to a process is checked.

Project procedures **must** be readily available to every project member as they form the common rules by which everyone works and, as procedures are issued or changed, then **all** project members affected by the procedure must be formally told of the changes.

Changes in project requirements may in many cases require changes to procedures and the process for making those changes itself needs to be documented.

Getting the best from people

Let us remind ourselves that this chapter is about project management, and management is about the efficient and successful handling of resources, the most volatile and sensitive of which are the humans we work with.

People issues are one of the most important areas of project management and probably as much has been written about this as any other subject. These notes only scratch the surface – locate some of the good texts and videos which are available on this complex area. Careful management and utilisation of human resources is the root of a successful and timely project.

The individual practitioner and the corporate manager will need the same basic people skills, but a different emphasis on how they are implemented.

Each manager will need to select those areas which meet specific needs. For instance the individual practitioner may well feel that to have a well-honed sales technique is close to the top of the list of required skills. And maybe the corporate project manager will consider that interviewing skills, to ensure the right people are on the team, will be a major priority.

Today, the development of people is a well-recognised requirement and is very high on the skills list of managers. Appropriate courses which managers may wish to consider cover such subjects as: listening and feedback, contract management; closing a sale and dealing with poor performance.

One particular area where some development will bring rewards is in the appropriate management of meetings – meetings usually involve a team of people (whether it is a team brought together for the meeting or a long-standing team) who are meeting to review their progress or a technical problem.

Meetings

Different kinds of meetings need different treatments, but at the root of a successful meeting is good planning and meeting management. The elements to be considered when planning a meeting include:

- Determining what kind of meeting is to take place, for example, review, problem solving, planning, decision making, information, general team meeting
- Understanding the outcomes you want to achieve from the meeting, for example, what decisions should be made, what problem is to be solved, what information is to be imparted, what change is to be achieved
- Ensuring that the appropriate people are present, that is, those who can provide input to assist in achieving the outcome or those who have the problem to be solved
- Publishing the agenda (well in advance of the meeting) indicating how long will be allowed for each agenda item and, if necessary, the person to take the lead in the discussion of that item
- Handling the meeting (the process you will adopt to ensure that the outcomes are achieved). Note that the meeting is not 'yours' – as chairman you are the facilitator to get the meeting called, underway and proceeding on the right lines – you muster and guide the participants to the outcomes
- Being aware of the way different people behave and react in meetings and making use of these different behaviours to the benefit of the meeting, at the same time ensuring everyone has an opportunity to contribute.

During the conduct of the meeting itself a whole range of skills needs to be exercised to ensure that the energy and resources brought together create a cost-effective result. In particular, the chairman needs to summarise discussions, clearly define actions, elicit acceptance and responsibility for actions and obtain agreement on timescales for their completion.

Following the meeting, the discussions, agreements, actions, completion dates and other relevant facts need to be included in the minutes, to unambiguously record any outcomes, actions and other important information. It is generally frowned upon to assign actions to non-participants of meetings, because they have not had an opportunity to provide their input or agreement. Assigning actions to a meeting member who subsequently delegates that action to another person is a far more effective and equitable method of achieving a more effective result. Those who have actions must feel that they own them and follow them through.

At subsequent meetings, actions are generally reviewed for progress and any further activity – actions are the building blocks for project progress.

One final thought – meetings can be cancelled as well as arranged. The costs of meetings are significant (just work out the cost of people's time!!). Only go ahead with a meeting when it is sensible to do so – not just because 'we always have this meeting at 2.00pm on a Wednesday'.

Working smarter

Encouragement of one's colleagues and our personal motivation to review the methods and processes involved with our day to day work, as a normal part of our activities, can be a major factor in the successful management of a project. If we are to deliver a product on time and to specification whilst minimising costs we must continually review the way we perform work.

Continuous improvement is a popular term for this activity where individually, or in informal or more formal groups, team members review their work processes and deliverables, looking for new methods of performing a task and eliminating any wasteful process steps. Again, there are many well written texts on this subject, and lots of tools (including management consultants) to help you come to decisions on the ways to proceed or what to change.

Don't forget, 'if you always do what you have always done - you always get what you have always got'.

Teamwork

However you look at this particular subject, whatever your industry viewpoint, the documentation professional relies on a team to help them through. The team will either be a combination of one or more subcontractors who pass on a developing product from one to another or a multi-functional team of documentation specialists who work together delivering their individual outputs from one to another for further processing.

Teams can be developed which provide a completely integrated set of the skills required to fulfil a task. Teams are formed to complete a specific task and are split up – they may exist for a few hours or for years on end. What is known is that an effective team performs far in excess of the sum of its parts.

One major process that the Project Manager can perform within the team is to work to combine the strengths of the individuals to reach the ultimate goal, motivating the team to accept that goal as their own. Successful teams need to have a common vision and be consistent in their promotion and drive to make it happen.

Project execution

So far, we have covered preparing for and winning work and arming ourselves with some skills that will generally assist in project management. Plans have been developed to show how the project will progress, how it will be staffed and how costs will be disbursed throughout the life of the project.

These are the major tools to be used during project execution. The plans are continually referred to and progress monitored against those plans. If circumstances change the plans can be amended to provide the new goals, milestones and objectives. The project team should commit to the new objectives and the Project Manager and other supervisory staff should pass consistent messages to the whole team to ensure common combined efforts.

Many of the thoughts and comments made above are encapsulated in *The Project Manager's 10 Commandments* below, first published in the early 80s by a South American company, and used and modified to suit individual requirements many times. They are a good *aide memoir*, applicable to both the sole practitioner and corporate manager alike.

Another useful reference, *Ten Things a Project Manager Should Do,* is a good document to use to carry out regular checks, to ensure all is well.

The journey ahead

Managing a project is just like captaining a sailing vessel on an arduous journey. The captain selects and organises the crew. According to the size of the vessel, the captain has many mates who can be relied on to pass messages to the crew, keep discipline and get jobs done – or on the smaller vessel the captain may be alone or have a minimal crew, and so will have to carry out most tasks personally.

The captain commences the journey on time, keeps an eye on the weather, watching for storms, ensures that the sails are set to get the maximum power from the prevailing conditions, and plots and alters course and sail settings as weather conditions change. The captain also regularly checks below decks to ensure that there is no infection or mutiny (and deals with these appropriately), and continually checks that the destination can still be reached on schedule as the journey progresses, using charts, tide tables and knowledge of the weather and the area in to help achieve this.

With all these jobs to do the captain needs to be able to interpret the signs and spot what is going wrong before the vessel run onto the rocks!

Check out the lists on the next two pages, and…

Happy sailing!

The Project Manager's ten commandments

1. Project Managers are **goal oriented**, knowing specifically what is wanted and what they are prepared to do to get it
2. Project Managers have **broad vision** and should adopt a holistic approach, recognising that there is far more to a project than the individual tasks and the logic connecting them
3. Project Managers should be **team builders** creating synergy within a group of individuals, harmonising and directing their energies to the common goal
4. Project Managers should have **technology and process understanding**, being aware of the existing and emerging tools and methods for performing tasks, but also recognising that the technology is secondary to the project tasks and the people performing the work
5. Project Managers should continually review and **optimise the project plan**, recognising new information and the effect that this has on the present position of the project and the changes needed to achieve the anticipated goal
6. Project Managers should be **flexible** in their behaviour and encourage flexibility in other members of the project team
7. Project Managers must **know their people and how they behave**, and utilise new innovations in behavioural science to develop the thinking and actions of the project team
8. Project Managers should **use conflict** to the good of the project, channelling the energy and emotions involved into achieving constructive work
9. Project Managers should be skilled in the techniques of **utilising individual and group creativity**, know when to use them and understand the dynamics that lead to releasing this skill
10. Project Managers above all should know what is **reality**, continually questioning the conflicting messages received and adjusting the perceptions of the project team - intuition plays a legitimate and significant part in determining reality.

Ten things a Project Manager should do

1. Issue a written Project Plan that has the explicit agreement of both management and the project team.
2. Make sure that each estimating duration is realistic in view of the resources allocated to it and is accepted as a target by the task manager.
3. Set up a communications strategy with all people on or involved with the project and ensure that it is adhered to.
4. Monitor the progress of each task against the plan at regular and other appropriate intervals.
5. Listen to the project team to understand the real problems on the project.
6. Recognise the **real** external constraints under which management is operating, even when this calls for changes in the project plan.
7. Monitor costs and resources consumed against the agreed plan.
8. Check that deliverables from each task are as they should be in terms of quality, timeliness and quantity.
9. Ensure that the quality standards are laid down for each task and that the quality control procedures are built into the project plan.
10. Continually review the project for the opportunity to innovate and not be afraid to make changes if they make sense.

14
How to write a synopsis

Roy Handley, MISTC *is working as a Technical Author after serving 24 years as an Aircraft Technician in the British Army. Employed primarily in the aerospace/defence industry, he was awarded 1st place in the 1999 ISTC Class B documentation awards for authoring a user guide for a Spectrum Analyser.*

Introduction

A definition of a synopsis defines it as meaning an outline or general survey of a subject. When compiling a synopsis the author should always remain fully cognisant of the definition. A well thought-out and constructed synopsis will prove an invaluable tool in the preparation of a document, whether it is in the design, planning or estimating stage.

In its design a synopsis can take any form that the author wishes. However, the content is normally suited to the type of document that the synopsis has been prepared for. Example 1 illustrates a synopsis for an Air Publication for the Ministry of Defence.

Example 1

Brief

The brief has defined the requirement as: to prepare to Camera Ready Copy an Air Publication for the Mk1 Parachute Assembly. The publication, which will comprise Topics 1, 2 and 3, shall include:

(i) Introduction
(ii) Maintenance
(iii) Packing Instructions
(iv) General Repair (First Line)
(v) Preservation and Storage
(vi) Modifications
(vii) Spares
(viii) Repairs (Third Line).

The publication is to be written in accordance with AvP7O (Specifications for Air Technical Publications).

From the initial information supplied, basic construction of the synopsis can commence.

SYNOPSIS 1
Mk 1 Parachute Assembly

HEADING	TOPIC/CHAPTER/PARA	ILLUSTRATIONS
N/A	PRELIMINARY PAGES	
	Page (i)/(ii) Front Cover	NO
	Page (iii) Amendment Record	NO
	Page (iv) Record of AIL/STAL	NO
	Page (v) Contents	NO
	Page (vi) Modification Record	NO
General and Technical Information	TOPIC 1	
	Chapter 1 Introduction	YES
	Chapter 2 Maintenance	NO
	Chapter 3 Packing Instructions	YES
	Chapter 4 Repair	YES
	Chapter 5 Preservation and Storage	NO
General Order and Modifications	TOPIC 2	
	Page (i) Front Page (Title)	NO
	Page (ii) Preface	NO
	Page 1/2 Equipment Modification List	NO
Illustrated Parts List and Related Information	TOPIC 3	
	Page (i)/(ii) Front Page (Title)	NO
	Page (iii)/(iv) Modification Record	NO
	Page (v)(vi) Preface	NO
	Page 1 Index of Part Numbers and Detailed Parts List	NO
	Page 2 Parachute Assembly	YES
	Page 3/4 Pack & Harness	YES

So, from the brief, a basic synopsis has been produced in order to indicate to the author the broad content of the document. The next phase would be to take each part of the document and draft a synopsis that designates the specific topic requirements. The next example illustrates a synopsis for Topic 1 Chapters 1, 2 and 3 for the Mk 1 Parachute Assembly.

Example 2

SYNOPSIS 2
Mk 1 Parachute Assembly
TOPIC 1

HEADING	TOPIC/CHAPTER/PARA	ILLUSTRATIONS
Topic 1		
Chapter 1	Introduction:	
	Leading Particulars	NIL
	General Description	YES
	Parachute	YES
	Harness	YES
	Pack	YES
Chapter 1 Remarks:	**Introduction to include Table listing component parts of assembly. Table to include Part No. NSN Nomenclature. Units per assembly.**	
Chapter 2	Maintenance:	
	Anti-deterioration	NIL
	Scheduled	NIL
	To be written in tabular form under the following heading:	
	Anti-Deterioration Maintenance:	NIL
	Preparation	NIL
	Examinations	NIL
	Completion	NIL
	Scheduled Maintenance:	NIL
	Before Issue:	NIL
	Preparation	NIL
	Dismantling	NIL
	Cleaning	NIL
	Examinations	NIL
	Testing	NIL
	Assembling	NIL
	Completion	NIL
	After Use and Periodic Maintenance:	
	Preparation	NIL
	Dismantling	NIL
	Disposal	NIL
	Cleaning	NIL
	Examination	NIL
	Assembling	NIL
	Completion	NIL
Chapter 2 Remarks:	**Safety and Maintenance requirements shall be covered under Safety and Maintenance Notes, which shall precede Anti-Deterioration Maintenance.**	

As the two examples illustrate, the synopsis has developed from the brief to identifying the chapters. It literally is a 'living' document that can and frequently does change as the project develops, although the first synopsis should establish the guidelines.

In constructing a synopsis the author should not feel bound by any style guides. It can be as brief or as complex as the project particulars dictate.

Preparation

A synopsis is generally presented in two forms:

- Narrative
- Tabular.

Narrative

A narrative synopsis would normally be produced for a small document such as a Technical Leaflet or a Sales Leaflet. It may include

- Product
- Type of finish, that is, glossy or matt
- Illustrations
- Content (the brief).

Advantages of a narrative synopsis are that it is simple and easily understood. Disadvantages are that it can be too simplistic and therefore deflects from the subject.

Tabular

A synopsis laid out in tabular form is ideally suited for the larger documents, for example Operators Handbook, Maintenance Manuals, Spare Parts Catalogues. The advantages of a synopsis prepared in tabular form are:

- Logical sequence
- Clear definition.

There is only one real disadvantage of a synopsis produced in tabular form and that is its preparation time.

As stated earlier the format of a synopsis is at the author's discretion. For illustrative purposes we shall continue to develop the Parachute synopsis. Example 3 covers Chapter 1; timescales are included.

Example 3

SYNOPSIS 3
Mk 1 Parachute Assembly
CHAPTER 1

HEADING	TOPIC/CHAPTER/PARA	ILLUSTRATIONS
Introduction	Leading Particulars: Ref. No NSN (Complete Assy) Part No Dimensions and Weight Length Width Depth Weight: Dry, Wet	 NIL NIL NIL NIL NIL NIL NIL NIL
Introduction Remarks: NSN required. Time-scale 0.5hr		
TABLE 1	Table listing component parts of Parachute Assembly (Details from GA drawing) Table to include: NSN, Nomenclature, Part No. Qty	NIL
Table 1 Remarks: NSNs to be applied for. Time-scale 1-0hr		
General Description	Parachute Assembly (2 para) Parachute (1 para) Harness (1 para) Pack (1 para)	1* 1** 1** 1**
General Description Remarks: **1. Parachute Assembly description to include all sub-assy interface** **2. 1* to be B/W photo of deployed assy.** **1** to be orthographic B/W** **Time-scale 4.0hrs (excluding illustrations)**		

So from the three examples above it can be seen how the synopsis develops. Remember it is up to the author as to how much information is included. A typical initial synopsis is illustrated below.

One factor to consider for inclusion in the synopsis is the timescale, which is beneficial for estimating purposes. Authors nearly always underestimate when estimating. There are no hard and fast rules, although as a general guide allow one hour per page to write the first draft, although it is down to the author's individual skill level. When providing timescales consider the following:

- Deliverables (CD, floppy, hard copy)
- Research
- Project plan
- Synopsis
- First draft (proofreading etc.)
- Illustrations
- Vetting
- Changes to first draft
- Second draft
- Validation
- Final changes
- Prepare to CRC
- Print/bind/collation
- Distribution.

In conclusion, a synopsis should always be included in the design stage in order to provide a summary for estimating and planning. If it is well thought out it should clearly define the document, from chapter to illustrations, diagrams, tables and so on.

When combined with a program of work, a synopsis could define the entire structure of the project. Once a synopsis has been prepared it should be formally approved, including any subsequent changes, before commencing the project.

Typical synopsis

JOB NO 1234

DESCRIPTION CABIN SMOKE DETECTOR SYSTEM

TOPIC/VOLUME/CAT/CHAP/SECT/PARA (Delete as applicable)			
REQUIREMENTS	**REMARKS**	**HRS**	**PAGES**
Chapter One Introduction Description + Figs Operation + Figs	 1 Para 4 Paras 3 Paras	 ¼ 2 1	 ½ 1½ 1
Chapter Two Maintenance	 Operator level only. Tabular	5	6
Chapter Three Removal and Installation + Figs	Detailed procedures to include tools and materials	25	20/30
Chapter Four Repair + Figs	Repair up to LRU level only	15	10/15
	TOTAL	50	50

Note: Author time only

15
The challenges of new media

Brian Gillett, MISTC *is Sales and Marketing Director of Author Services Technical (AST). Brian specialises in designing Technical Publications solutions which balance the needs of individual clients and their end users. He advises on the use and implementation of existing and new media as the tools to support these solutions. This role requires him to keep pace with new media and to continually review and develop the solutions in light of new media.*

Introduction

As technical communicators our core skill of taking complex technical information and presenting it in a format comprehensible to the end user has not changed over time. What has changed is the increasing variety of media on which to present the information. Our challenge is to keep pace with these changing media.

Whether we are part of a large corporation or a personal service company, the need to be aware of both the advantages and, equally importantly, the disadvantages of new media should be of great importance to us. It is in all our interests to increase the recognition of our industry by continually improving the way we deliver our information to our end users.

I have described some of my thoughts on how best to meet the challenges of new media. Each of us will have peculiarities about the environment we work in, yet the way we meet the challenges will be the same. The formality of approach to the challenge will be very much dependent on the size and complexity of the media.

What do we mean by new media?

As the word media can mean different things to different people I think it important to define my meaning of the word in order to put my thoughts and suggestions into context. The Collins English Dictionary gives the definition of media as 'a means or agency for communicating or diffusing information'. I think of media, in the context of technical communication, as the means for compiling and delivering information to the end user.

I refer to new media as being new *to you* rather than the very latest media. What is a new media to one person may be a familiar media to another.

New media can take many forms but can generally be broken down into five different categories:

- Environment
- Transmission
- Formats
- Hardware
- Software.

Environment

By this I mean the actual route of delivery of information. The hardware and software for delivery will usually already be defined so it is unlikely that as technical communicators we would be involved in the development of the environments. Our challenge is to understand the impact of how the delivery of information in different environments affects our end users, and what limitations are imposed on us for the delivery of the information.

Transmission

Communications technology is developing rapidly and costs are coming down dramatically, opening up opportunities for more and more of us to use these new media to transfer information to our end users. We are likely to use only one of these communications technologies so the challenge is to **review** the best option for our needs and the needs of our end users.

Formats

New formats are being developed to enable information to be delivered and viewed in the new environments. It is unlikely that we would use all of these as a solution to improve our generation and delivery of information, so the challenge here is to **review** which best suits our needs and the needs of our end users.

Hardware

New hardware is coming onto the market on a regular basis. We may use this as a means to archive information as well as deliver it to our end users. Our challenge is to **review** which hardware we need to meet our chosen solution.

Software

Software tools to format information are many and varied. Once we have decided on the format to deliver our information we then need a software tool to produce the information in that format. Many software tools profess to do the same thing, but each will do it in a different way and, where no industry standard exists, may do some things better than others.

What are the drivers?

The need to face the challenges of new media will be dependent on your individual circumstances, but the underlying principle is that you will be facing the challenge because it either offers cost savings to your company or provides a better product for your end users. In many cases it is likely that both your company and the end user will benefit. Whatever the circumstances the adage of 'if it is not broken do not try to fix it' no longer applies as business today is conducted in a 'how can we improve it' culture, essential to survival in competitive markets.

If you are not continually striving to improve your efficiency in generating and delivering information then you could be left behind.

A large technical publications department can lose its internal business to an outside company that can demonstrate cost savings simply by using new media. A personal service company may lose the next contract to another personal service company that can do the work quicker by using better tools or win the business by offering a better solution using new media.

There are usually many reasons why we have to face these challenges. Typically these may be summarised as:

- Client driven
- Market driven
- Business driven.

Client driven

Some customers look to us, the technical communicators, to produce their documentation on the media that is most appropriate for the job. In this case, as consultants, it is up to us to seek continuously to improve the service we provide. The challenge here is to continually **review** new media.

Other customers, internal as well as external, are aware of what media are available on which to generate and distribute information and hence they will dictate to us the media to be used for their work. If the media required is not one we are familiar with then the challenge is to **implement** the use of the media.

Market driven

The market place, in which our companies operate, continually changes. Our competitors always seek for better ways to work. Certain industries revolve around large corporations who set the standards within their industries, thus dictating the market. Smaller companies in the same industry are often forced to follow the market lead set by large corporations. The challenge then is to **implement** the new media within your own company.

Business driven

In some cases it may be that you need to introduce new media in to your working practices as the result of a business merger or directive from director level. It is likely that business driven initiatives are given to standardise media across the company, and it is not likely to be a decision in which you can participate. The challenge here is to **implement** the new media in to your working practices.

What approach can we use?

There are three general ways to meet the challenges of new media:

- Outsourcing
- Delegation
- Teamwork.

Outsourcing

Outsourcing the challenge can often be a very effective solution. The knowledge of the company you outsource to will save a lot of time and effort by yourself or your colleagues. In addition the focused expertise of the company can often provide innovative solutions to meet your needs. When choosing a company to outsource to, look for a company that uses a diverse range of media already and can demonstrate understanding of exactly what you require.

Delegation

If you are in a position whereby you are able to call on colleagues to carry out the review on your behalf, this may be a viable way of carrying out the review – it is an opportunity to use the existing knowledge of colleagues. In some circumstances this may be your only option due to company policies not to outsource or a lack of resources to put together a team to review and implement the new media.

Teamwork

For new media, which will have an impact on different people within your company, it is recommended to put together a team of people who would be affected by the new media. The team may include members of marketing, IT, Quality, Sales and so on. Each representative will be able to review the impact on their own departments and would enable an informed opinion to be made on the effectiveness of the new media from a company-wide perspective.

Process

Whichever way you decide to meet the challenge the process will be the same. There are two key stages to the challenge of new media:

- Review, which can be summarised as:
  Investigate
  Assess
- Report.

Review

Investigate

There are several good ways to investigate new media:

- Test it
  Ask manufacturers to provide you with evaluation and beta copies to test. In many instances they may well be more than willing to demonstrate the product to you. Create a case study to test the product.
- Speak to someone who is using it
  If you have been active in the technical communications industry you have probably already met someone who is using the product. The adage of 'everyone likes to be asked advice' is one you should use.
- Gather information about it
  Press releases, product brochures, exhibitions and information on the internet are good sources of information, and readily available. Over the internet you can subscribe to numerous free news bulletins, which will keep you informed of new media.
- See it in action
  Attend exhibitions where you can see the product in action or use your networking powers and find someone who is using the media to demonstrate it to you.

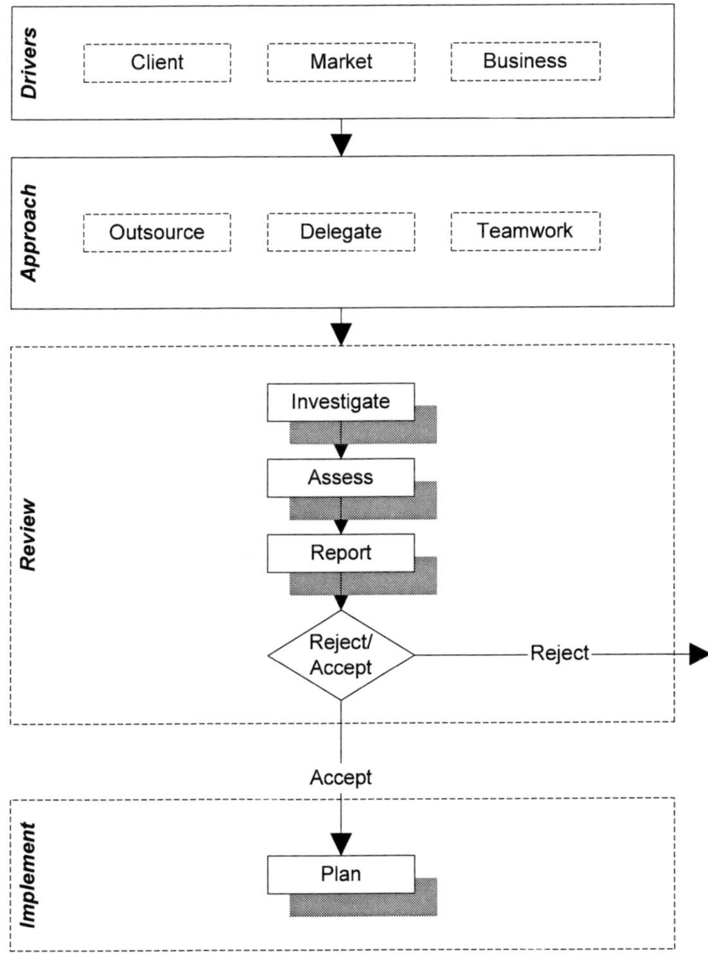
Drivers
Client
Market
Business
Approach
Outsource
Delegate
Teamwork
Review
Investigate
Assess
Report
Reject/
Accept
Reject
Accept
Implement
Plan

Assess

The overriding feature to look for is – 'will the new media be better than the current media for both you and your end users?'. Questions you should ask yourself may typically include:

- Benefits to you
  - Will this give your company a competitive edge over your rivals?
  - Does the new media enable you to produce a product which is better than your competitor with little or no additional cost?
  - Will the media improve timescales – will the media save you time?
  - Will the media give added value to your customers – does the media have a perceived value to it which your customer would appreciate?

- Costs to your company
  - Can it be introduced easily?
    Availability of the media will affect when you can be up and running. Look out for media that requires other new media for it to work that will add extra costs.
  - Can personnel be trained?
    Courses provided by the supplier will enable your staff to be productive quickly but look out for expensive training courses that are mandatory with certain media.
  - What are the delivery costs?
    Hidden costs may be incurred when delivering the information, so review fully.
  - Do I need any additional software to use the media?
    The new media may require the use of additional software that will incur additional costs.
  - Do I need additional hardware to use the product?
    The new media may require the use of additional hardware that will incur additional costs.
  - Is support given to the media?
    Free telephone or on-site support as well as upgrades may be important things for ease of implementation and to ensure additional costs are not incurred in the future.

  - Is it simple to use?
    Simplicity will reduce the cost of training.

- Benefits to your end user
  - Is it the industry standard?
    Will your end users be familiar and conversant with the media? Industry standards are likely to be familiar to your end users whereas non-industry standards are not.
  - Is the media proven?
    If many other companies are using the media then this is likely to be an indication that it is a proven one and that your end users will accept the media. Find out if the media is being used by companies your end users will know.

- Costs to your end user
  - Is the viewing software free?
    There may be free plug-ins or readers to enable you to generate the information in the full product and then deliver it to your customers at no further cost to yourself or your customers.
  - Do your end users have to pay to view the information?
    What does the end user have to purchase and how much is it?

- General considerations
  - Does the media have longevity?
    Find out what the manufacturer's plans are for the media (upgrades and so on). If they are planning other non-compatible media to replace it then you should be wary.
  - Do I have a choice of products?
    Are there several versions of the media so that you can select the one which best meets your needs? Some media come with everything, which can be expensive, whilst media that come in different 'flavours' are often cheaper than the full-blown media and easier to use.

- Is the media available now?
  Is the media still under development or is it available now? Be wary when looking at media which are under development, as promises of usability and functionality are not always met.
- What are the future enhancements to the media going to be?
  What plans does the manufacturer have for the media or compatible enhancement media?
- How accessible is the media?
  Is the media portable, easy to move around and transport for exhibitions and so on?
- Is the media reliable?
  Talk to others to see if they have had problems with the media.
- Is the media similar to other media already in use?
  Similar media to those already in use will be easier to adapt to quickly due to an acquired familiarity.
- Is the media global?
  If you need to deliver your information to other subsidiary companies or to overseas customers the availability of global media becomes an important feature.
- Do other companies support the media with plug-ins?
  Some software products license third party companies to develop additional functions to work with the media. This is a good indication that the media is proven and has longevity.
- Will your customers accept the media?
  Tell your customers that you are in the process of reviewing a new media and request their thoughts. This discussion may well have a big impact on whether to go ahead and implement a new media. Your customer may even have already carried out some initial reviews of the media, which they may be willing to share with you.

Report

Having completed the other two stages it is time to document your findings. This serves several purposes but the key ones are:

- To give a record of findings which you can share with others
- To help to sell your case when seeking budgets to go ahead and implement

The report may be as formal or informal as needed but bear in mind that the report can be used as a basis for all future reviews and provide a comparison document against future media.

The report may typically include:

- Summary of findings
- Identification of the impact on end users and internal staff
- Review of the balance of benefits vs. costs
- Recommendation (reject or accept)

Implementation

Once you have decided that a particular new media is the right one for your company then you should be committed to implementing it. Some media may need a long time to implement, meaning that during the implementation stage another new media may appear on the market that can do the job equally well or better. When you make your choice ensure that you have reviewed it fully and go with it. If we all waited for that next new media 'just around the corner' then we would never change, and remember change is an integral part of our business function and means we stay competitive.

To implement new media efficiently you need to define a plan of action that addresses responsibilities and rollout timescales for the new media to be in place. Training requirements defined in the review should be carefully planned to ensure that staff are able to work with the new media effectively. Do not forget any training requirements for your end users.

How can we make the challenge less daunting?

The challenges of new media are rooted in the availability of time to review and implement them. Many people see it as a burdensome task fraught with unknowns. The reality is a lot less daunting if we have a plan of action and a method of meeting the challenge.

I suggest that we should all strive to be proactive in meeting the challenges of new media, whatever role we play in the business. We should put into place mechanisms to continually evaluate the way we produce and distribute information to our end users. The question to ask ourselves is: 'is there a better way to do this', to reduce our effort to produce and distribute the information and to provide a better product for our end users.

Summary

In summary we should adopt an attitude of willingness to review new media. Any decision that the media is right for our requirement will be made easier and confidence will be gained to implement it. Care should be taken to carry out reviews professionally. Bear in mind all the elements that must be addressed to ensure that the new media is right for our company, and our end users, and do not just accept the next release of the media as the magic solution we have been looking for.

16

Determining your audience

Colin Battson, FISTC *began his career as an Aircraft Antenna Design Engineer with Standard Telephones & Cables Ltd, following a student apprenticeship. Seven years later he embarked on a second career in technical writing, beginning as a contract author on-site with Marconi Space Division. In the 30 years or so since then, his work has embraced a wide range of disciplines and topics, whilst employed for the most part by Technical Publications Consultancy companies. He is presently Chief Author for Author Services Technical Ltd, in Letchworth, Herts.*

How this chapter is structured

What is a 'target audience'?

This section explains the meaning of 'target audience'; the readers of your document or publication, whether paper or on-line.

Why you need to identify the target audience

This section explains why it is important, not just to the audience (the readers and/or users) but to you as the author of the work.

When should you research the audience?

This section explains why it is important and most cost-effective to do your research as soon as possible, preferably before a word has been written.

How to determine the audience

This section summarises some of the common ways of achieving your goal, and compares the feasibility of the different methods in different circumstances.

How knowing the audience affects your work

This section describes the advantages in cost-savings, productivity and other more subtle ways, that writing targeted material can bring.

The effects of missing the target!

This section covers some of the problems that can be caused by a poorly-targeted piece of work.

Summary

This section is a round-up of the key points covered.

What is a 'target audience'?

The generally-accepted meaning of the term 'target audience' is the typical reader or user of the work being produced or to be produced.

Although it is possible that anyone at all may read the subject material, it is important to target the style and level of the content at the anticipated majority audience group.

Taking two extreme examples of publications and associated audiences:

1. A first reading book for (say) five year old schoolchildren: This would typically contain lots of colour to make it visually appealing, it would probably have a picture to complement every word or short group of words, and it would comprise a fairly small number of pages (short attention span)
2. A User Guide for an application software package: Although colour might be used, many such User Guides are black/white only. Language and

terminology would be aimed at a computer-literate person; perhaps someone assumed to have a basic understanding of Windows techniques. It is likely the document would be of the order of 50 to 150 pages, sub-divided into chapters, and undoubtedly would include features such as Contents List, Glossary, Index, etc. It could also include one or more Appendices containing reference material.

These two examples demonstrate two very different target audiences. However, it is often necessary to define the audience to a much tighter definition. Factors which might require consideration for Example 2 above could include:

- Is the reader assumed to be familiar with the subject or related software (e.g. if the User Guide covers an update or enhanced software version)?
- Is the subject publication part of a suite of publications? If this is the case, it might be that more detailed or specialised information would be most appropriately included in an existing Reference Manual or Programmer's Manual, for example
- Will the reader have any supporting information to back up the subject publication, e.g. On-line Help files?
- Will the reader have had the benefit of formal training on the subject software?
- Has the software been deliberately designed to require a relatively low level of skill/expertise from the user? If so, the User Guide should reflect this philosophy.

For every document or publication, the target audience can and should be determined if the reader / user is to obtain maximum benefit from it.

Why you need to identify the target audience

What are the primary reasons for determining the anticipated target audience for a proposed new document or publication?

There are clearly advantages to be gained, both for the users and for the author(s), if the document produced is aimed precisely at the correct target audience.

Advantages to the users:

- Easily understood, hence readable and of value
- Not too complex (confusing) or too basic (boring)
- Will be considered as useful, and therefore more likely to be used
- Will help them avoid errors caused by lack of understanding.

Advantages to the author(s):

- Avoid additional work putting in surplus detail
- Avoid re-work following submission of draft
- Enhancement to reputation by 'getting it right first time'
- Reduced costs/timescales arising from all the above
- Increased job satisfaction, especially following good feedback from users.

When should you research the audience?

The right time to do your research is **as early as possible**!

Because it is so important to determine the audience level in order to set the depth and scope of the content, it is clearly better to have that information before a word is written.

You may not always enjoy the luxury of that situation (for example, work may already have been started by someone else before your involvement; perhaps engineers have produced a draft, and so on).

However, if you are able to do the research at a very early stage, that is most definitely the time to do it.

If you are producing the subject document(s) or publication(s) for a client company, your costed quotation could be significantly affected by having taken into account the target audience specification. Indeed, it could be a major factor in whether you or your company secure the work at all.

> **HINT**: If you are tendering for the work in competition with others, do ensure that your target audience definition is included in your quotation. It will help your prospective client compare 'like with like' and will show them that you have actively considered the target audience in your document design.

If you don't do your research early enough, you could be facing major comments on your draft submission, failure to meet project deadlines, plus potential loss of reputation and future business as a consequence. On top of all that, if the badly or untargeted document does get as far as publication and distribution to end users, you and your client are likely to incur the wrath of the users once the inadequacies of the document are realised.

Remember that it's not just the style and depth of treatment that have to be addressed; it is also important to ensure consistent use of approved terminology. Continuing with the software User Guide theme, use of the terms 'window' and 'dialog' should be consistent. Similarly, is data entered into a 'field' in a window, or into a 'dialog box'? Does the user 'select' something in a window or 'click on it'?

Attention to this level of detail is part of audience determination, and should not be neglected when researching your audience. Get hold of existing documents whenever possible, and solicit comments on those from your client. You need to know if the existing material is considered to be satisfactory. If it is not, ensure that you take on board any general criticisms which could affect the approach to the planned new document(s).

In the case of software publications, a good User Guide can save the company a great deal in Help Desk and similar support costs. The converse is also true. Although targeting the audience correctly is just one of many factors involved, its importance in that context is clear.

How to determine the audience

You know that you need to determine the target audience, and you know that you need to do that as soon as possible, but how is it actually done?

There are various methods employed; not all will fit every situation, and sometimes that intangible called 'experience' plays a major role.

However, for those who perhaps don't have the depth of experience needed, here are some of the methods used:

Informal:

- Discussion with your client
- Interviews with designers/engineers
- Interviews with end-user company/individual users (where possible)
- Comparison with related, existing publications - preferably published by/for the same company.

Formal:

- Obtain a copy of the governing House Style manual or Style Guide
- Follow the governing publication specification, if applicable
- Prepare a questionnaire for completion by end users and others.

The more of the above you are able to use, the more precise will be your target audience determination. You may well find that you will obtain conflicting information as you build up your 'picture' of the typical end user. That is to be expected and should be resolved before work commences.

Different individuals - even those in the same category (e.g. end users) - are bound to have preferences not shared by all. Your goal is to review all the information you obtain, then to finalise the requirements by further discussion - probably with your client, but perhaps also with a nominated representative of the end users.

Whatever happens, make every effort to define and document the agreed audience parameters <u>before</u> writing is started; re-work is a waste of time and money!

How knowing the audience affects your work

Having made the effort and invested the time involved in determining the target audience, you should be ready to start preparation of the subject document(s), ideally with the results of your audience determination activities formally agreed with all interested parties.

If you are in this position (and you should be!), then you can begin writing with some degree of confidence, and can expect that the draft(s) will:

- Be understood by the end users
- Not be too simple in approach, nor too complex
- Attract minimal comment regarding style, depth and terminology
- Not be returned with major re-writes required.

If you are able to achieve this pleasurable state of affairs, the results are likely to be:

- You will meet or improve on target delivery date(s)
- Your reputation will be enhanced
- You are likely to secure repeat business
- The end users will use the documents and regard them as having real value
- The level and cost of support activities (Help Desk and so on) will be reduced.

The effects of missing the target!

If you don't properly research the audience requirements, the consequent problems are likely to be the negation of all the points listed as benefits in the previous section.

Your own job satisfaction is guaranteed to be reduced to an all-time low ('How can I have got it so wrong?') and you probably won't be offered any further work from the same source.

That sums up your side of the issue, how about the client and end-users?

- The job will no doubt have run over time and budget, due to re-write(s) activity needed

- The client will be frustrated and lack confidence in your work, possibly criticising every little detail that otherwise might have been overlooked
- The end users will probably not want to use the publication (especially if it was published in its 'inadequate' form). As a consequence, Help Desk and other support costs will escalate
- Possible damage or destruction of product due to misuse by uninformed users
- In extreme cases, possible litigation could arise.

Summary

Important key points to remember:

- The advantages to both author and user of properly targeting the audience
- Do your research as early as possible for maximum benefits
- Choose the best ways to determine your audience for each project; use as many methods as possible
- There are many enduring benefits to getting it right!
- Get it wrong (or don't do it at all), and you may not get a second opportunity.

Appendix 1

Some useful British Standards

The list below specifies some of the most relevant British Standards.

For more information see the British Standards Web site at www.bsi.org.uk. For information about other relevant standards including aviation industry standards see the ISTC web page at www.istc.org.uk.

Alphabetical arrangement and filing order **BS 1749**

Bibliographic referencing: electronic documents **BSISO 690-2**

Book spines **BS 6738**

Citing and referencing published material **BS 5605**

Computer software management guidelines **BSISO 9294**

Determining subject matter for indexes **BS 6529**

Documentation for computer based systems **BS 5515**

Documentation for users of application software **BS 7649**

Forms design and layout **BS 5537**

Glossary of documentation terms **BS 5408, DD247**

Indexing **BSISO999**

Loose leaf binders **BS 5097**

Loose leaf publications **BS 5641**

Marks for copy preparation and proof correction **BS 5261 pt2**

On-screen documentation for users of application software **BS 7830**

Pages, sizes **BS 1413**

Paper sizes **BS 4000**

Point numbering **BS 5848**

Preparation of copy for microfilming **BS 5444**

Preparation of typescript copy for printing **BS 5261 pt1**

Quality manuals **BS 5750**

Research and development reports **BS 4811**

Rules for drafting and presenting safety standards **BSEN 414**

SGML **BS 7138**

Signs **BS 559**

Tables, graphs and charts **BS 7581**

Technical drawing **BS 308**

Technical manuals **BS 4884**

Titles **BS 4148**

User documentation **BS 7137**

User's requirements for technical manuals **BS 4899**

Visual aids for lectures **PD6482**

Appendix 2
The ISTC web site

The ISTC web site is a very useful resource for professional communicators. Its contents include:

- Articles from the Communicator magazine
- Information about the ISTC conference and conference papers
- Book reviews
- Information about education
- Resources and standards

The site also has valuable links to related sites.

Find out more by visiting www.istc.org.uk

Appendix 3
Proofreading marks

This appendix shows the proofreading marks that should be used to mark up copy and to correct printers' proofs, as specified by BS 5261. They are classified into three groups:

- General
- Deletion, insertion and substitution
- Positioning and spacing.

For each marking up or proof correction instruction a distinct mark should be made in the text, to indicate the exact place to which the instruction refers, and in the margin, to signify or amplify the meaning of the instruction. Note that some instructions have a combined textual and marginal mark.

Classified list of marks

Note: The letters M and P in the notes column indicate marks for marking up copy and correcting proofs respectively.

Group A - General

Number	Instruction	Textual Mark	Marginal Mark	Notes
A1	Correction is concluded	None	/	P Make after each correction
A2	Leave unchanged	- - - - - - - under characters to remain	(✓)	M P
A3	Remove extraneous marks	Encircle marks to be removed	╳	P e.g. film or paper edges visible between lines on proofs
A3.1	Push down risen spacing material	Encircle blemish	⊥	P
A4	Refer to appropriate authority on anything of doubtful accuracy	Encircle word(s) affected	(?)	P

Group B - Deletion, insertion and substitution

Number	Instruction	Textual Mark	Marginal Mark	Notes
B1	Insert in text the matter indicated in the margin	⋏	New matter followed by ⋏	MP Identical to B2
B2	Insert additional matter identified by a letter in a diamond	⋏	⋏ followed by for example ◇A	MP The relevant section of the copy should be supplied with the corresponding letter marked on it in a diamond, e.g. ◇A
B3	Delete			MP
B4	Delete and close up	through character or through characters, e.g. charaacter characacter		MP
B5	Substitute character or substitute part of one or more words	/ through character or ⊢⊣ through word	New character or new word(s)	MP

Number	Instruction	Textual Mark	Marginal Mark	Notes
B6	Wrong font. Replace with correct font	Encircle character(s) to be changed	⊗	P
B7	Set in or change to italic	——— under character(s) to be set or changed	$\underline{///}$	M P Where space does not permit textual marks encircle the affected area instead
B8	Set in or change to capital letters	≡ under character(s) to be set or changed	≡	
B9	Set in or change to small capital letters	═ under character(s) to be set or changed	=	
B9.1	Set in or change to capital letters for initial letters and small capital letters for the rest of the words	≡ under initial letters and ═ under rest of word(s)	≡	
B10	Set in or change to bold type	∿∿∿∿ under character(s) to be set or changed	∿	
B11	Set in or change to bold italic type	$\overline{\sim\sim\sim}$ under character(s) to be set or changed	$\underline{///}$ ∿	
B12	Change capital letters to lower case letters	Encircle character(s) to be changed	≢	P For use when B5 is inappropriate
B12.1	Change small capital letters to lower case letters	Encircle character(s) to be changed	≠	

Number	Instruction	Textual Mark	Marginal Mark	Notes
B13	Change italic to upright type	Encircle character(s) to be changed	⊥	P
B14	Invert type	Encircle character to be inverted	↻	P
B15	Substitute or insert character in 'superior' position	/ through character or ⋌ where required	⅂ under character e.g. 2 over ⅂	P
B16	Substitute or insert character in 'inferior' position	/ through character or ⋌ where required	L over character e.g. L over 2	P
B17	Substitute ligature for separate letters	⊢—⊣ through characters affected	⁀ e.g. ﬃ	P
B17.1	Substitute separate letters for ligature	⊢—⊣	Write out separate letters	P
B18	Substitute or insert full stop or decimal point	/ through character or ⋌ where required	⊙	MP
B18.1	Substitute or insert colon	/ through character or ⋌ where required	(:)	MP

Number	Instruction	Textual Mark	Marginal Mark	Notes
B18.2	Substitute or insert semi-colon	/ through character or ⋏ where required	;	MP
B18.3	Substitute or insert comma	/ through character or ⋏ where required	,	MP
B18.4	Substitute or insert apostrophe	/ through character or ⋏ where required	'	MP
B18.5	Substitute or insert single quotation marks	/ through character or ⋏ where required	' '	MP
B18.6	Substitute or insert double quotation marks	/ through character or ⋏ where required	" "	MP
B19	Substitute or insert ellipsis	/ through character or ⋏ where required	...	MP

Number	Instruction	Textual Mark	Marginal Mark	Notes
B20	Substitute or insert leader dots	/ through character or ⋏ where required	(…)	MP
B21	Substitute or insert hyphen	/ through character or ⋏ where required	|–|	MP
B22	Substitute or insert rule	/ through character or ⋏ where required	|—|	MP Give the size of the rule in the marginal mark, e.g. |1em| |4em|
B23	Substitute or insert oblique	/ through character or ⋏ where required	(/)	MP

Group C - Positioning and spacing

Number	Instruction	Textual Mark	Marginal Mark	Notes
C1	Start new paragraph			MP
C2	Run on (no new paragraph)			MP
C3	Transpose characters or words	between characters or words, numbered if necessary		MP
C4	Transpose a number of characters or words	3 2 1	1 2 3	MP To be used when the sequence cannot be clearly indicated by the use of C3. The vertical strikes are made through the characters or words to be transposed and numbered in the correct sequence
C5	Transpose lines			MP
C6	Transpose a number of lines	3 2 1		P To be used when the sequence cannot be clearly indicated by C5. Rules extend from the margin into the text with each line to be transposed numbered in the correct sequence

Number	Instruction	Textual Mark	Marginal Mark	Notes
C7	Centre	enclosing matter to be centred		MP
C8	Indent			P Give the amounts of the indent in the marginal mark
C9	Cancel indent			P
C10	Set line justified to specified measure	and/or		P Give the exact dimensions when necessary
C11	Set column justified to specified measure			MP Give the exact dimensions when necessary
C12	Move matter specified distance to the right	enclosing matter to be moved to the right		P Give the exact dimensions when necessary
C13	Move matter specified distance to the left	enclosing matter to be moved to the left		P Give the exact dimensions when necessary
C14	Take over characters, words or line to next line, column or page			P The textual mark surrounds the matter to be taken over and extends into the margin
C15	Take back characters, words or line to previous line, column or page			P The textual mark surrounds the matter to be taken back and extends into the margin

Number	Instruction	Textual Mark	Marginal Mark	Notes
C16	Raise matter	over matter to be raised under matter to be lowered		P Give the exact dimensions when necessary. (Use C28 for insertion of space between lines or paragraphs in text)
C17	Lower matter	over matter to be lowered under matter to be lowered		P Give the exact dimensions when necessary. (Use C29 for reduction of space between lines or paragraphs in text)
C18	Move matter to position indicated	Enclose matter to be moved and indicate new position		P Give the exact dimensions when necessary
C19	Correct vertical alignment			P
C20	Correct horizontal alignment	Single line above and below misaligned matter, e.g. misaligned		P The marginal mark is placed level with the head and foot of the relevant line
C21	Close up. Delete space between characters or words	linking characters		MP
C22	Insert space between characters	between characters affected		MP Give the size of the space to be inserted when necessary

C23	Insert space between words	between words affected		MP Give the size of the space to be inserted when necessary
C24	Reduce space between characters	between characters affected		MP Give the amount by which the space is to be reduced when necessary
C25	Reduce space between words	between words affected		MP Give the amount by which the space is to be reduced when necessary
C26	Make space appear equal between characters or words	between characters or words affected		MP
C27	Close up to normal interline spacing	(each side of column linking lines)		MP The textual marks extend into the margin
C28	Insert space between lines or paragraphs	or		MP The marginal mark extends between the lines of text. Give the size of the space to be inserted when necessary
C29	Reduce space between lines or paragraphs	or		MP The marginal mark extends between the lines of text. Give the amount by which the space is to be reduced when necessary

Index

Compiled by INDEXING SPECIALISTS, 202 Church Road, Hove, East Sussex BN3 2DJ. Tel: 01274 738299. email richardr@indexing.co.uk.

Note: **Bold** page numbers refer to chapters.